Praise for *The Words of My Father*

"An urgent, impassioned call for peace between Palestine and Israel. . . . Eloquent and affecting memoir. . . . An inspiration to peace activists in all theaters of war and struggle and a book that deserves a wide audience."

—*Kirkus Reviews*

"Bashir's childhood in Gaza and his life under occupation meld in this retelling, and his charming earnestness shines through. This moving tribute to Bashir's remarkable father is also a compelling argument for peace."

—*Library Journal*

"[A] candid and deeply felt coming-of-age story. . . . Even in the face of great adversity, Bashir prevails as an optimistic champion of peace. . . . This moving meditation of a young man's struggle to find peace amid turmoil will resonate with readers concerned with Israeli-Palestinian relations."

—*Publishers Weekly*

"At one of the darkest times in the elusive effort to end the conflict in the Middle East . . . [comes a] remarkable new memoir . . . full of youthful exuberance, unlikely adventures, and raw discovery. . . . Captivating."

—*The New Yorker*

"Beautifully written by a young Palestinian Growing up in the Gaza Strip, *The Words of My Father* is a deeply personal narrative imbued with a clarity that surprises ᵃᵗ . . . . The sheer humanity of his story maᵈ

ry Supplement

"A work of profound spiritual beauty, one of the great memoirs to emerge from this terrible conflict. . . . *The Words of My Father* offers all of us hope that this seemingly intractable conflict can find a solution that is just to both sides."

—Yossi Klein Halevi, author of the *New York Times* bestseller *Letters to My Palestinian Neighbor*

"*The Words of my Father* is a book about loss—the loss of land and dignity for Palestinians living in the occupied territories, the increasing loss of humanity among many of the occupiers, and the loss of the author's beloved father who spent most of his life preaching peace, but tragically died in the midst of increasing polarization between Palestinians and Israelis. By making us *feel* the plight of the Palestinians from the inside, this poignant and beautifully written book should inspire us all to raise our voices for change."

—Robert Greene, *New York Times* bestselling author of *The Laws of Human Nature* and *The 48 Laws of Power*

"In a year that has seen a relentless focus on life and death in the Gaza Strip, Yousef Bashir's memoir is a moving reminder of the normality of Palestinian aspirations in what has been an abnormal situation for far too long. This book is a remarkable testimony to overcoming hatred and fear."

—Ian Black, author of *Enemies and Neighbours: Arabs and Jews in Palestine and Israel, 1917–2017*

"To experience love and humanity on many levels, read this story. Beautifully told by a young man whose voice deserves to be heard—even if the world is not yet ready to listen."

—Diana Darke, author of *My House in Damascus: An Inside View of the Syrian Crisis*

The

WORDS

of My

FATHER

The
WORDS
of My
FATHER

———— ✦ ————

Love and Pain in Palestine

YOUSEF
BASHIR

HARPER ⬤ PERENNIAL

NEW YORK • LONDON • TORONTO • SYDNEY • NEW DELHI • AUCKLAND

HARPER ● PERENNIAL

Originally published in the United Kingdom in 2018
by Haus Publishing.

A hardcover edition of this book was published in 2019 by
HarperCollins Publishers.

HarperCollins books may be purchased for educational,
business, or sales promotional use. For information, please
email the Special Markets Department at
SPsales@harpercollins.com.

FIRST HARPER PERENNIAL EDITION PUBLISHED 2020.

Map by Martin Lubikowski

Library of Congress Cataloging-in-Publication Data
has been applied for.

ISBN 978-0-06-291733-1 (pbk.)

20 21 22 23 24   LSC   10 9 8 7 6 5 4 3 2 1

To my fellow Palestinians
struggling for peace, justice, and freedom.

"We must not let our wounded memory guide our future."
—Khalil Bashir

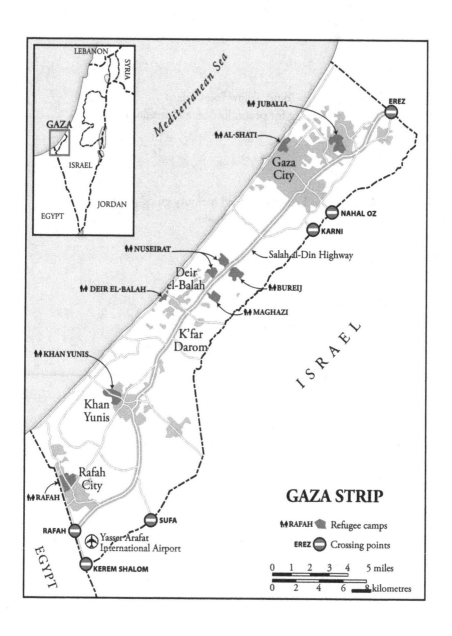

LEBANON

SYRIA

Mediterranean Sea

GAZA

ISRAEL

JORDAN

EGYPT

🕌 JUBALIA

🕌 AL-SHATI

Gaza
City

EREZ

NAHAL OZ

KARNI

🕌 NUSEIRAT

Salah al-Din Highway

Deir
el-Balah

🕌 DEIR EL-BALAH

🕌 BUREIJ

🕌 MAGHAZI

K'far
Darom

🕌 KHAN YUNIS

I S R A E L

Khan
Yunis

Rafah
City

🕌 RAFAH

RAFAH

SUFA

✈ Yasser Arafat
International Airport

KEREM SHALOM

EGYPT

GAZA STRIP

🕌 RAFAH 🟥 Refugee camps

EREZ 🔘 Crossing points

0   1   2   3   4   5 miles

0   2   4   6   8 kilometres

Contents

The
WORDS
of My
FATHER

Before I Start

I have learned to read my audiences. If there is a hush as I am led onto the stage, then I know I have already hooked them. If, however, the buzz in the room continues while someone shows me my seat and adjusts the microphone, I know they are thinking about other things and I am going to have to work a little harder.

If the room is small, like a meeting room or a classroom, I start working on eye contact right away. I suppose it is my way of announcing myself. In a large hotel function room like this one, I just scan the crowd. Some look back at me, as curious about me as I am about them. Most are checking their iPhones for important messages from friends about what they had for lunch. Others are networking. A typical Washington crowd. If I were not on the podium, I would be doing the same.

They are well dressed. These three-piece-suit people will ask questions that are heavy with facts and are more statements than questions. Students in jeans and T-shirts usually ask broader, more general questions.

An older woman comes over to me with a glass of water and makes sure I have everything I need. I do. She reviews the session's format. Introduction. Speak for half an hour. She will give me a signal when my time is almost over. Twenty minutes of questions. I smile and thank her. I do not say that the questions will run long after the twenty minutes have ended, that when the session is over there will be a knot of people who come to the stage to ask even more and stay there long after everyone else has left the hall, or that a small group will trail me out to the lobby until they head off one by one. This happens all the time.

I look out over the audience. There are a couple hundred of them.

I presume that many, if not most, are Jews; this event is hosted by the American Israel Public Affairs Committee, which more than any other entity impacts US policy toward Palestine. These people have voices that get heard in Washington. It is important that they hear me.

One of them will ask me about the Israeli settlements on Palestinian land. I have an answer for that, but perhaps they will hear me differently after I tell them my story.

"Tell them your story, Yousef," I can hear my father saying. "Tell them your story, then at least they will know something they did not know before."

My story is in so many ways my father's story. My grandfather's story. And the story of so many grandfathers before him. For centuries before Europe exploded and sent its Jewish people in search of safety, my family had farmed the land along the Mediterranean coast in Gaza. Our town, Deir el-Balah, is named "Monastery of the Date Palms" after a monastery established there by the Christians in the 4th century CE. Even before then, it was already an important stopping point along the route to Egypt. It once had a fortress. Alexander the Great passed through, as did the Romans, the Ottomans, and probably even Joseph and his brothers.

My family, the Bashirs, watched them come and go, and maybe even sold them some of our dates. Two things had brought my ancestors to this place: fresh water springs that never ran dry, and the sweetest, reddest dates on the Mediterranean coast. We have been there forever. And for the past three hundred years since records were first kept, our name has been on them for this land.

Though my father, Khalil Bashir, was known as an English teacher in the local schools and later as a high school headmaster, our land was so much a part of his identity that he thought of himself as a farmer.

"This land has been in my family for hundreds of years," I heard him say many times. "It is my childhood. It is my memories. It is my family. I cannot leave, because history is watching and I am not prepared to make the same mistake my people made in May 1948 when they evacuated their homeland. This is a lesson to everybody: you must never give up your land or your country, otherwise you will lose your dignity and your life."

A well-dressed young man steps up to the podium and quiets the room. The audience puts away their iPhones, at least for the moment. I will know that I am losing their attention if the phones reappear.

From the moment the young man says that I am a Palestinian, the room becomes extra quiet. It always does. I look out and smile. I can almost hear them thinking, "But he looks so Jewish." Several, in fact, will say that to me afterward. I just want to tell them now, "Of course I look Jewish. We all have the same genes. The scientists have proven it. We are cousins. So why all these problems?"

The young man briefly recounts my bio and mentions that I hold an MA from Brandeis University. "What kind of Palestinian goes to Brandeis?" some of them are thinking. I know this because I get that question all the time. I welcome their curiosity. It will make them more interested in what I have to say.

It is my time to speak. There is polite applause as I stand. For a split second, I am not sure where to begin. This always happens when I am facing a big group. I have brought some notes, which I will not actually use, but I look at them on the podium for a moment and let my father's voice fill my mind.

"Tell them your story, Yousef," I hear him say. I will do that. But first I will tell his.

"I want to start by telling you two things about my father," I say. I speak quietly. "My father was as much in love with his land as he was with my mother. And he loved both of them deeply."

Part One

PARADISE

"Believers, eat the good things we provided you, and
thank God, if you are truly worshipping Him."
—The Holy Quran, Surat Al-Bakarah (2:172)

1

Our Land, Our Neighbors

My father was as much in love with his land as he was with my mother. And he loved both of them deeply.

He loved the tall date palms along the edges of our farm that marked out an area larger than two football pitches. He loved his olive trees and the oil he produced from them. He loved his hives both for the bees and the amazing honey they produced pollinating his guavas, figs, and dates. He especially loved his greenhouses—long tunnels of plastic stretched over metal frames. They covered almost half the land. He raised tomatoes in them, along with hot peppers and aubergines that he sold to markets in Israel and Jordan, and sometimes Egypt. He had spent his lifetime enriching the sandy soil with compost to make it immensely productive.

He loved how on almost every piece of the land around his house there was a tree. Some, like his figs and guavas, gave fruit; others brought shade and color to the open areas around our house. There had once been an orange grove, but my father's brother had cut down most of the trees to make space for a new house he built for his family. My father never completely forgave him. He loved the memory of those orange trees, which he still saw in his mind, where they had stood for generations.

He had more than two hundred date palm trees. Some grew around the corners of our house while others lined the long driveway from the house to the gate onto Mekka Street. There was a tower for pigeons, a cage for many rabbits, geese, ducks, chickens, and at least two roosters. We tied our gentle old white donkey by a small room which housed the motor that powered the irrigation system and our water well. Next to it was a very large olive tree, a couple of hundred years old. Olive trees can live for centuries. I liked to climb it to find a good hiding spot where I could relax and scrutinize my universe.

We had dozens and dozens of olive trees, and during olive-picking season we all turned into a team. For breakfast, my mother fried eggs mixed with salted potatoes and cheese, and baked bread in her mud oven. My father spread plastic beneath each tree so that any olives that fell could be picked up without getting covered in the dry earth. I enjoyed climbing the trees. The higher I got, the more excited I felt. I came down only when my mother announced that lunch was ready: jasmine rice with beef and okra stew. She served it on a tray with a smashed tomato salad and a plate filled with freshly cut tomatoes, cucumbers, and hot peppers (mostly for my father), all straight from our farms, along with pickled olives from the previous season.

Even though I found the whole process exciting, by the second day I was looking for ways to finish faster. One way was to use a stick to shake down the branches without having to pick them by hand. My grandmother got upset with me for this because she did not want the branches to be scratched. She lectured me about how an olive tree is like a human being: its skin is not to be harmed. I just waited for her to get distracted and went back to using my stick.

At the end of the land some distance from the house, my two older brothers had helped my father build a hut from where we could keep watch on the far side of the property. It was a stockade of dry palm trunks with woven palm leaves for a roof. It had a place to make a fire and lay long, thin cushions called *freash* on the floor, where we sat.

Our land was so big, I never ran out of space to be on my own. It was a paradise. Knowing that I grew up in Gaza, you may be surprised to hear me call it "paradise." Today—and my heart breaks to say so—you might be more inclined to call it "hell." Believe me, though, when I was a child it was not like that, and I still think of it the way it was then.

When the harvest came and the date clusters were red and heavy, my father's crews of pickers climbed the trees for weeks, filling trucks that left every day for the markets and packing houses. He saved the best ones for my mother to make her amazing date jam. That was the finest gift we could give anyone, and I always gave a jar to my teachers when I needed to make sure I was on a sports team.

My mother grew up in the city. She had been born in Beirut, where her family owned land, but came back to Gaza at an early age. Her uncle had been the mayor of Deir el-Balah for many years. Her family is one of four original families there. They owned a lot of land. Some of her relatives were uneasy about her marrying my father. They saw themselves as cosmopolitan city people and were dismissive of him for being a farmer and less well off, though the Bashirs are the largest of the four original families. But my mother's mother, Sitie, liked him. He was handsome, with movie-star good looks. He was immaculately dressed, even when he was working on the farm. When he spoke with somebody he looked directly at them; his dark eyes filled with such sincerity that his listeners were left deeply touched. He tilted his head just slightly as he spoke, chose his words carefully, and treated everyone kindly.

"Leave it with me," Sitie told my father. He did. Eventually, he and my mother did marry. Over the next dozen years, they had eight children. I was the fourth, with two older brothers and an older sister, plus one younger brother, then a twin sister and brother, and finally a youngest sister.

He built a house for my mother and made it from the highest quality cement bricks, which looked like stones cut from a quarry. It was a sharp contrast to the humble house that still stood next to it, where he, his brother, and two sisters had been raised by their parents. He remodeled one of the rooms into a garage for his white Opel, and used it as a place to keep some of his farm tools. We stashed large bales of hay in another one of the rooms to feed the animals and birds we raised.

For many years, the new house had had only one story. In 1997, when I was eight, he started building two more floors using the money he had made from his greenhouses. He and my mother saved their money carefully. They did not believe in taking out a bank loan. They did something only when they had the money to pay for it, and were confident it would remain theirs until Judgment Day.

The work on the new floors kept getting delayed, however, because our neighbors took my parents to court to stop the construction. The neighbors were Jewish settlers and Israeli soldiers living illegally next to and on our land. The two new upper floors would make our house the tallest building in the area. Our neighbors did not want that. On many occasions, my parents left early in the morning for Tel Aviv, across the Israeli border, only to come back from the court late at night empty-handed. In the end, my parents won their case, but it took two years.

A Jewish settlement had been built directly across the Salah al-Din Highway from our house. It was called K'far Darom. Two earlier settlements had been located there, one in the 1930s and another a decade later. In both cases, the settlers had left after only a few years. Following Israel's occupation of Gaza in 1967, a third settlement was established in 1989—the year I was born.

When I looked over at the settlement from the upper floors of our house, it was so unlike everything else in Gaza. Their houses were laid out along two sides of a paved road that almost made a circle at the center of their two hundred and fifty acres. Many of the houses had satellite dishes. Their roofs were sloping and covered with red clay tiles. My mother told me that was how houses were built in Europe.

Though K'far Darom looked beautiful, the settlers themselves gave me an uncomfortable feeling every time I saw them. They just did not seem friendly. I had watched on TV how violent settlers had been toward Palestinians, how they showed no regard for children or women. Those images hung at the back of my mind when I saw the men always working with small machine guns strapped around their bodies. They scared me even though I had never seen them fire their weapons.

I never saw children, only older people and women in long black dresses with their heads covered in black turban-like hats. We had heard that the residents of K'far Darom were among the most religiously conservative of the eight thousand illegal Israeli settlers in Gaza.

They had greenhouses like ours, but hired Filipinos to work in them. They built a cement footbridge across the Salah al-Din Highway—not far from our house—to connect their homes on the far side to their synagogue, which was on our side. The footbridge was open only to the settlers.

Between our land and the synagogue was an Israeli military base. It had two high watchtowers, where soldiers with machine guns were stationed twenty-four seven. The towers were about forty feet tall, just slightly lower than our house. They were made of metal girders, and each had a small room at the top. We could see one of them from our formal living room and the kitchen, and the other from our bathroom and my parents' bedroom.

The soldiers stationed in the towers had binoculars. Whenever we looked at the towers we could see the soldiers watching us. Because they were always there, we presumed we were always being watched.

Despite this, the soldiers on the base for some reason were less intimidating to me than the settlers—maybe because they always seemed to be enjoying life more. Every Saturday the soldiers turned their music all the way up; most of their music sounded very Arabic to me. We heard them as they clapped and cheered to the Egyptian pop singer Amr Diab's *Habibi ya Noor El Ein* [*Baby, You Are the Light of My Eyes*].

They often played football and sometimes kicked their military-looking footballs across our fence. Sometimes, they came and asked for their ball back, though that was rare. Sometimes I kicked their ball back on my own. Sometimes I played with it a little before kicking it back. Sometimes I just kept it, because the palm trees had thorns that punctured my own balls. A new one was always welcome.

One of the watchtowers had been built right above the shed where we kept our animals. I thought that was funny because the roosters crowed so loudly that any soldier posted in that tower would have had to hate his life. Sometimes a soldier would decide that he'd had enough early-morning crowing and start jeering in Hebrew at the roosters, before he shut all the windows in the tower and disappeared along with his agitation.

"When the settlers first came," my mother told me one afternoon as she was making one of her tasty meals, "it was just us and the settlement. There was no base. No soldiers. The settlers had their guns, but we did not pay much attention to them. They went to buy fruit and vegetables from the market and the same shops where we bought our food." The Jews hated the Arabs, but they still went to Arab shops. The Arabs hated the Jews, but they still sold to them. I found that a bit ironic, even humorous.

I liked working with her in the kitchen. I was a foodie even then and was curious how she made things. Plus, it was a good place to talk.

"After the UN built their school just over there in front of our gate," she told me another time, "the students sometimes snuck into our greenhouses and grabbed cucumbers to eat on the way home. Your father used to go do some farm work near the road when their classes ended, to keep them away. One day while he was digging green onions, a settler parked his car on Mekka Street for some reason and the students started harassing him. The settler decided that Khalil was encouraging them. So, the next thing you know, he was aiming his gun at your father and forcing him into his car. For some reason, he made Khalil take off his shoes and leave them by the side of the road.

"I was in the house. I didn't see anything, but you know Mariam who lives next to the school, she saw what had happened and came running to tell me. What was I supposed to do? I didn't know what to do or who to call. How can a man be taken at gunpoint from his own land?

"Luckily, some soldiers in an Israeli military jeep had spotted the incident and drove after the settler. They rescued Khalil, but instead of bringing him home, they took him to the military base. They let the settler go, of course.

"The soldiers gave him problems about not having his ID card with him. He tried to explain that he was wearing his old jeans to work in the field and was not carrying his wallet. So, I had to go to the base and bring him his card."

Gaza has been under direct Israeli military rule since 1967. All Palestinians are issued an ID card at sixteen, and our personal information is stored on Israeli computers. When the soldiers checked my father's ID and found he was "clean," they told him he could go home.

"No apology. Just, 'Go home.' That was all." My mother put down the potato she was peeling and broke into a big smile. "Khalil looked at them and said, 'I am a respected schoolteacher. I cannot afford to be seen walking to my home barefoot.' They thought that was funny and gave us a ride back to the house. They never returned his ID, though, so he had to apply for a new one."

It is in kitchens across Palestine that stories like these are told and remembered.

* * *

My older sister told me how the Israeli army base was built after the settlers attacked our house. That happened in 1992 when I was only three years old, and I have only the slightest memory of it.

"A young rabbi from the settlement—wearing an Uzi as always – was crossing the Salah al-Din Highway on foot. This was before the footbridge had been built. He was heading to the agricultural institute next to the synagogue where he worked. A young Palestinian attacked him with a knife in his back and pierced the rabbi's heart. The settlers wanted revenge for the dead rabbi. And whose house was closest?

"I was sitting on the veranda reading when one of the neighbors came shouting up the driveway, 'The settlers are coming. The settlers are coming.' Father was at school. Mother and Grandmother Zana raced around hiding their gold jewelry and the ownership deeds for the land and similarly valuable things.

"First, the settlers tore down the greenhouses, then set them on fire and burned our wheat fields and part of the orange grove. There were only a couple of them at first, but soon there must have been fifteen.

"Mother locked the veranda door and herded everybody into the living room, then locked the living-room door, which did no good. They broke down the veranda door and barged in. One of them picked up the TV—he really had to struggle, because it was heavy—then heaved it onto the floor and smashed it. Meanwhile, the others were breaking the windows from the veranda into the living room. They started spraying some kind of gas through the broken windows to force us to come out. We were all screaming."

I have a clear memory of that, but was too young to make sense of what was happening.

My sister went on. "A minute later, they broke down the door into the living room and began shooting at the ceilings and destroying so many of our things. The whole time, they kept shouting insults about Arabs.

"Then one of them grabbed my mother and told her to turn her head, because, 'I don't want to hit you on your stomach.' Can you believe he said that? He could see that she was pregnant with the twins. He did not want to hurt the babies, still he slammed her head with his pistol and knocked her down." When my sister got to that point in the story, she just started shaking her head. It was too much to tell anymore.

My courageous oldest brother, Yazid, who was only eleven at the time, tried to defend my mother. The settlers were grown men. They slammed him into the wall and punched him in the face, breaking his front teeth. The whole time they shouted at him, "This is for when you get bigger and throw stones at us!" Throwing stones was something my brother had never done and would never think of doing. He was the best student in his school and is now a successful orthopedic surgeon. Truth be told, I never thought throwing stones was a bad thing because I thought that the soldiers deserved it. They were doing much worse things to Palestinians than throwing stones.

When our Palestinian neighbors spotted my father speeding home to help us, they stopped him and tried to make him stay with them. They were afraid he might be killed if he went home, but he came anyway. He took us to the home of one of our neighbors, where we settled in for three days. Then, on his own, he went back to our house and remained there so the Israelis could not claim that we had abandoned it. In the meantime, there was a lot of chaos in town. There were riots in Deir el-Balah when several settlers came with a bulldozer and tried to knock down buildings. The settlers rammed the gate of the UN school before the Israeli soldiers made them stop.

By the time things were calmer and we felt it was safe to go home, the soldiers had cut down some of our orange trees and seized part of our land so they could expand their base and give the settlers more space on our side of the highway.

"The settlers attacked us, so they brought their soldiers to protect themselves," my mother always said sarcastically. She pushed my father to take the settlers to court. He agreed, but they had to sue in an Israeli court and, like so many other Palestinians, they lost in the end. They got nothing from Israel.

* * *

I was young when this happened and barely remember it, but I vividly recall the day we were all watching an Egyptian black-and-white movie and my father announced, "Put on your shoes and follow me. Right away!"

It turned out that the soldiers were digging a hole on our property without my father's permission and he was outraged. He ordered us all to climb down into the hole and stay there. I had no idea what this was all about. All I remember was how excited I was to be allowed to jump into such a big hole. I was staring right at the claw of the ugly-looking bulldozer not having a clue why it was there or where it had come from.

As my father jumped in, he very quietly but firmly declared to the soldiers, "If you want to keep digging you can do so, but you will have to dig over me and my children."

It was always exciting when my father went face to face with the soldiers. Even though they had guns, they could never get their way with him. For one thing, he spoke better English than most of them. My father had always been a good neighbor to everyone, so he expected others to treat him in the same way. He would not allow anyone, though, not even the soldiers, to take what was his. In the end, the soldiers retreated and we went back to our house.

That night I asked him, "Who are the *Yahood*?"

"*El Yahood*"—the Arabic word for Jews—was the term everyone used to refer to the soldiers. For some reason, the settlers were always known just as the settlers.

"They came from Europe," my brother Yazid volunteered. "I saw it on TV." But I wanted to hear about them from my father.

"They are our cousins," my father said.

"Our cousins?!" I replied with my eyes wide open. I guessed that explained why they always listened to Amr Diab's Arabic music.

"What kind of soldiers stay up all night and dance?" my mother complained to my father as he sat down with my grandmother to drink his tea. Around sunset, he liked to go to the veranda, chop some wood and light a fire in his *kanoon*, a small brass brazier that looked like a low table. We often gathered around it while we watched the news or entertained guests. My father loved his tea made on that *kanoon*—that's for sure. In the quiet of the evening, we could hear the music from the towers.

"Go talk to them," my mother insisted. "Once they get started, they will be making noise all night. We need to sleep." She had been bugging him about the music for days. Finally, no matter how much he tried not to, my father got fed up and gave in.

"Here, I am going to put down my tea and I am going to go talk to them, *mashe* [OK]?" he told my mother in such an upset yet amusing tone. They hardly ever argued in front of us and, when they did, my father made sure that my mother was soon happy again. He did speak to the soldiers, but the music did not stop.

* * *

As a child, I was excited every time soldiers came around to our house, all dressed up in military gear. They carried weapons that I had only seen on TV or in American action films. When they came off their base, I rushed to follow them and get closer to see how they looked and what they were up to. There was something about them that captured my attention—their uniforms, their boots, their helmets, the way they talked. I would get upset if I missed one of their unannounced appearances.

My father zoomed in on me when he saw my fascination. In no uncertain terms, he reminded me, "You are from here. Understand? You are the owner of this land." I did not really understand what he meant until I was older.

It was not unusual for the soldiers to appear suddenly out of nowhere, and then to disappear just as quickly. I might even catch a glimpse of one standing silently on alert among the small palm trees behind our house. On our land.

Sometimes we saw our Palestinian soldiers on patrol with them, but our guys did not have the same sense of style. They only carried AK-47s and wore blue uniforms. I thought they should have been wearing camouflage suits like real soldiers. The Israelis had helmets; our guys had hats—which they wore like they were making sure it did not interfere with their hair gel.

After each encounter with the soldiers, my youngest sister, Zana, liked to jump up and down and sing:

The neighbors went back home,
The neighbors went back home.

Zana did not know that the soldiers had no home behind our house. They were soldiers occupying our land because they could; apparently, no one could stop them—not even the marvelous Arab nations, nor the mighty USA.

* * *

For a few years, things remained fairly quiet. We were living our lives, and the soldiers and settlers were living theirs. Despite our strained relationship, we were living in peace, or at least in my own mind I was living in peace.

I felt no need to fear the soldiers. With my parents nearby, fear could never find me. No *Yahood* or settlers could get to me. When my father woke me up, tickled me, or rubbed my cheek against the stubble on his, I told myself, "This is where I belong. I am safe here."

I was so happy that I would sometimes say to my mother, "I will always love you and *Aboi* [my father]." With her chin nestled in the palm of her hand, she responded in her teasing way, "We shall see. We'll see who you will love when your gorgeous queen comes along."

If I ever had any worries, I went to the red metal gate that separated our driveway from the street. When my father arrived, we opened half of it to let his car pass through. There was a small door cut into the gate, and I would open and close that little door over and over, because, for some reason, doing so always made me feel untouchable.

My conflicts with my father were then still in the future. He was very good at making me happy. Sometimes when I was protesting something he wanted me to do, he would cajole me with some famous Kazem ice cream. Despite playing hard to please, I always gave in and he chuckled as he watched me eat. Of course, as soon as

I had finished, I started negotiating to get more, but my father was too wise to let that happen.

Life was going by smoothly. I took everything for granted. I thought it would always be like this: breathing fresh, cool air with birds peeping all around, lying under the guava tree staring at the pure blue sky, and dreaming that some day I would see the rest of the world.

2

My Family

My special love was football. I was always collecting football stickers. Every time I bought a new sticker I could not wait to get home and put it in my album hoping that some day I would win something, even though that never happened. I was always checking to see how close I was to filling its pages. For me, completing an album of football stickers was like writing a book, but with pictures instead of words.

I begged my mother to install a satellite dish so that I could watch Real Madrid matches. She refused. When the World Cup in France was broadcast on our local television, however, I watched every minute I could. I was amazed by the French-Algerian player Zidane. The whole house was cheering for him and for France. We were hosting a couple of German exchange students at the time. They watched with us, but did not cheer for France. During halftime, we all ran out to the courtyard for a quick match, with me doing commentary as we played.

The kids in our neighborhood always played football in the street, since they had no space around their homes as I did. They ran around barefoot and said the bad words they had heard their elders using. They were not very interested in school, and my mother worried that they were a bad influence on my siblings and me: by playing with them, I would become less interested in my studies. Truth was, I liked having friends from all backgrounds. Anyone could be my friend, which is why I was popular in school. My mother told me not to play with the neighborhood kids, but whenever my parents were away for a few hours, I would sneak out and kick some balls with them. The minute I saw my father's car appearing at the edge of my sight, though, I would bail on my team and sprint back inside our gate.

Sometimes I tried to imitate those kids by playing barefoot inside our gate. My mother would scold me with a shocked face: "Yousef, where are those slippers I just bought for you?" She did not like money to be wasted. My father often ended up having to play doctor and pull a thorn from my foot. He would say, "Don't worry, I won't hurt you," and before I could panic he would have yanked it out.

My father made it painfully clear that he did not consider football a good use of my time. He preferred that I was reading good books. I tried following his advice but found that the books he suggested were dull. I did not dare tell him that, though. He claimed I was too lazy to read, but I just wanted to read what I wanted to read.

When anyone sat down to watch a movie or a TV series I could provide a complete report about what the actors did in real life. When we watched football matches, I knew almost everything there was to know about the players, and even the referees. This would impress my father for a few minutes, but soon I would hear him growling, "I wonder if you know as much about what's in your schoolbooks!"

I spent hours trying to persuade everyone in the family to play football with me. Sometimes I would play by myself and do my usual running commentary. Sometimes this would motivate somebody to join me, but mostly I would end up playing by myself. When they finally did come, it was the most fun. Sometimes my father jumped in and even my sisters, and we would play and play. I loved it when my family played football or volleyball together. It was my secret pleasure.

My father's secret pleasure was tickling me. He would corner me when I was least expecting it and say with a big laugh, "Well, who do we have here?" He kept tickling me until my stomach muscles began to hurt from laughing. My older siblings even held me down so he could tickle me more easily. I had to break away and run screaming from the room before he would stop, and then he laughed even harder.

———

Even worse than the tickling was when friends of my grandmother Zana came to visit. They immediately grabbed me and kissed me on both cheeks at least a hundred times. There was nothing I could do to stop it. My mother knew it drove me crazy, but she just laughed. When I ran to the bathroom to wash off all the kisses, I shouted back at her, "Next time, I will not come say 'Hi.'"

Everyone was always laughing in our house, except my grandmother Zana, who was almost eighty, and was very quiet. She was always the first to wake up and recite her morning prayers with my father. During the day, she kept busy by doing lots of simple errands around the house. Her main job was taking care of the animals and making sure that they were well fed and had water. This was a job she assigned herself. My father would have preferred that she do nothing and just make herself comfortable. She kept telling him, though, that she liked to keep busy. At night, she slept on a *freash* on the veranda with her medicine in a small, beautifully made wooden box that hung nearby on the wall.

Whenever my grandmother heard anyone criticize me, she quickly came to my defense.

"He is the one who always helps me when I need anything," she said. Of course, I then felt very guilty whenever I did not do something she asked me to do. She just made a gesture with her hands and said quietly, "I will pray for you." She was so thin and so light, I could pick her up like a paper.

She liked to tell me how when I was very small she had taken me to the orange grove to look for the owl in the daytime and the wild pigeons in their nests. I cannot really remember doing that. I loved it, though, when she took my hand to go with her somewhere.

* * *

I loved going anywhere with anybody and jumped at any excuse to make the three-mile trip into Deir. However, nothing was better than being in the car, riding with my parents when they went to

Gaza City. This happened at least once a week. My brothers, sisters and I had to wait for our turn to go with them. Going in the car was an important privilege granted by my parents. If I had gone the last time, somebody else would get to go the next time.

As we drove through the red metal gate and into Mekka Street, I felt a sense of adventure growing inside of me. I always wanted to know what was going on outside our world. I loved sticking my head out the Opel's window and embracing the pure winds of the sea. I wanted my father to do the same and feel it, too, but he was always too busy talking to my mother who sat beside him in the front seat.

The road was paved and wide enough for two lanes, but had no line painted down the center. As we drove, most of the houses we passed were owned by relatives—some close, some distant. You had to go a long way to find a house that did not know a Bashir.

As we approached the commercial heart of Deir, we would first come to the Bank of Palestine and a small supermarket where we bought Palestinian-made products like *laban* and Al Juneidi yogurt, which I loved, instead of Israeli ones. Further on were other markets: the local street market, where farmers sold from stalls and carts, as well as the bigger and fancier Abdul-Razek supermarket across from the Israeli military base. If we had kept going beyond the intersection at the center of Deir el-Balah, we would have driven through the refugee camp located by the beach and the coastal highway.

Many of the refugees had come from the villages and farms east of the Strip—places where their families had owned land for centuries. Now they were crowded together in small houses with no gardens or courtyards and no trees nearby. The Strip is less than eight miles across at its widest, and half as wide in most other places. It has very little space for all the Palestinians forced to move there. My grandmother Zana told me stories of how we and many other locals had opened our farms and houses in 1948 and in 1967 to feed and host the exiled Palestinians.

It took many years before I realized how blessed my family had been to have deep roots in Deir el-Balah. Our lives were so different from those who had come to the Strip as refugees, especially those who had to live in the camps.

* * *

As I grew older, I spent long hours trying to convince my mother that I was old enough to run errands in Deir for her by myself.

"I showed you that I know how to get to the shop and back to the house. You promised you'd let me go alone."

"You are too young, Yousef." That was all she said, for years. Then one day, while I was busy watching *Tom and Jerry*, I saw her shadowy reflection in the TV screen.

"Yousef, will you go to the market for me?" she asked.

I was so focused on the cartoon that I shouted without turning or looking at her, "Yes, yes, but not now."

"Not now?" she wondered aloud. "All right, then I will ask someone else."

No! That was a challenge. *Tom and Jerry* was suddenly unimportant. I raced after her into the hallway, shouting, "No, no! I can go now, I can go now."

Going to do errands for my mother soon became a regular task. Eventually everyone began to ask me to get them things. My older sister would come give me a hug and ask ever so nicely, "Yousef, will you go get me a Galaxy?"

Galaxies were her favorite chocolate bars. She would tell me what a kind brother I was, and it was not long before I was out the door. When I got back, it was always a different story. She would grab the Galaxy from my hand and run back to her room, shut the door, and go back to studying.

* * *

Everyone else in my family was always studying. If we were not reading or writing, we were reminded we were not making "good use

25

of our time." If I got ninety on a test, my father asked me in a miffed tone, "Why not ninety-nine or a hundred?" He would look at the test paper as if he would find the answer there, then say something like, "Singing, Ping-Pong, sports, and hair gel. I get it now."

My brothers studied the way my parents expected. I wanted to do things my way. I did not care if I got into trouble, and I did, many times. I wanted to enjoy the things that I could do living in such a beautiful place. I was much happier doing almost anything but the dreaded schoolwork. This distressed my father very much.

One place where no one pressured me to study was at the house of my mother's mother, Sitie, who was my favorite person. She lived in a two-story villa in Deir el-Balah close to the Salah al-Din Highway. Her husband, who had died before I knew him, came from wealth, and the house showed it. She had housekeepers, a garden full of mangoes and figs, and animals that she liked to look after. Four of her sons and their wives and children lived with her, so there were lots of cousins in the house. That made it fun when I visited.

Sitie was a very tall, matronly figure and very well respected in Deir. Whenever she went to the market all the shopkeepers wanted her to come into their shops. When I went shopping with her, she treated me as though I were her assistant, and I loved it.

I adored staying with her. She treated me like a prince. She would grill me chicken and pass it to me as I watched the Spacetoon channel. I savored the okra stew she cooked called *tabeekh*. She also made me *kishk* by letting yogurt dry until it was hard. Sometimes she gave me *mamoul*, a small round pastry with dates inside that she and the other women in the house had spent a long time preparing. Best of all, her TV had access to Real Madrid matches. She had a satellite dish because she liked to watch Al Jazeera.

She enjoyed having me sleep over at her house. Every time she came to our house, even if she had come for some serious family matter, she asked, "Can Yousef come home with me?" For some reason, though, my father did not like me staying at other people's houses, even at Sitie's. I never understood why.

"Your house is where you sleep," he stated firmly many times.

Sitie had a lot of respect for my father. Whenever her children quarreled with one another, she would call on him to mediate. Hence, she did not press him when he refused to let me stay with her, but she never hesitated to ask again the next time.

Once, when for some reason I was allowed to stay there for a few days, I saw that my uncle Mujeeb went to the sea every morning to go fishing. Some hours later he would return with the fish he had caught. Sitie was impressed, but I was suspicious. One morning he took me with him. After a couple of hours with no luck catching anything, he told me to hold the bait while he went and did some business at the fish market. On the way home, I began to be suspicious about why the empty bag we had brought with us had become so heavy. Of course, when he gave the fish to his mother, I just had to announce, "Sitie, he didn't catch them, he bought them at the market."

Uncle Mujeeb was outraged at having his secret revealed. He ran after me most of the way from Sitie's house to my house, shouting, "Stop! Stop right now." I never did. I was too fast for him, but that ended my visit with Sitie that day.

* * *

I never met my uncle Hassan, one of Sitie's sons. All I knew of him was that he had no fear of the unknown, liked adventure, and had died in New York City as a young man. According to my mother, I am a second version of him.

"You are as notorious as Hassan," she would tell me when she was upset with me. "You even look and act the way he did, and that is what worries me."

"Tell me more," I asked excitedly.

"About New York?" she answered with her eyebrows knit together.

"No! Not New York. Tell me about Uncle Hassan." But she always changed the subject.

I wanted to do things on my own, hoping I could prove to my parents that I was capable of being independent. I looked for ways to get my father to stop pressuring me to study. In spite of my resistance to my father, I had tremendous respect for him. On the surface he appeared to be easy-going, but the truth is he was always busy and expected his children to be as well.

Sometimes when I was playing video games (always Atari), he would get upset and ask, "Yousef, why don't you work on the land?"

Nonchalantly, I would answer, "No, not now." Even when I did go out to join the workers in the greenhouses, I always ended up slowing the process down so my father or the workers would ask me to leave. I did not mind. The greenhouses carefully controlled the air in a way that was good for the plants, but it felt heavy on my lungs and I was happy to get back outside.

Sometimes, after large tomato orders were shipped off, we had leftovers to sell in the local market. My father woke at 3:00 AM, said his prayers, hooked up the donkey and cart—he refused to give them up because they had belonged to his father—and off he went. One morning I went with him. I could tell my father was pleased that I was finally showing some interest in the farm.

I liked watching how he conducted business with the traders who gathered around our cart and tried to outbid each other in the funniest ways. I liked it even more when I was allowed to collect the money and keep the change.

I liked it so much that I was then always eager to go with him. Still, he asked the night before in a doubtful tone, "Yousef, you have to be up by 4:00 if you want to go. Can I count on you?"

I would shout back, "I'll be on time, Yaba."

Always, the image of a fresh falafel sandwich after we finished was lurking in my mind.

3

The School

In September 2000, my father became the headmaster of a new school in Deir. It was a gift to our city from a wealthy German named Rudolf Walther. My father insisted on naming the school after him. The two of them shared a deep commitment to peace, and both believed that education was its cornerstone. As they got to know one another, they became very close friends.

I never met him, but I heard that he had spent some time in prison during his youth. While there, Walther had promised God that, if he ever got out, he would work hard to earn money and would use that money to support the cause of peace. In addition to funding the school in Deir, he donated funds to construct other schools in Gaza, the West Bank, Israel, and numerous other places around the world.

My father was a natural choice to head the new school. For many years, he had been a teacher of English in the Sokynah High School for girls located next to where the Rudolf Walther School was built. In a traditional society such as ours, it was a mark of great respect for a man to be made a teacher at a girls' school. Partly this was because he came from one of the most highly regarded families in Deir; mostly, though, it was because people recognized him for the profoundly honorable man that he was.

He loved teaching, and he loved the English language. After many years in the classroom, though, he accepted a promotion to become the principal of a primary school. Then, only a year later, he was asked to lead the new Rudolf Walther School.

The school was an enormous white building. It was co-educational for the first six grades, but boys-only in the upper grades. The year that it opened, I went there for sixth grade. When the bell rang, we would all gather in a line and walk into class with the teachers at our side as if they were guards. My classroom on the third floor was right above the headmaster's office on the first floor.

I made friends quickly. One of them, Omar, shared a desk with me in the front row. Each desk was for two students, and each classroom usually had about forty to fifty students. Omar was very pale and quiet; his small round glasses made him look extremely innocent. The history teacher accused him of being like a snake, which was an accurate description. Omar came across as harmless, but he was in fact only closing in to bite.

In the desk right behind us sat two cousins, Hossam and Ibrahim. Hossam was at heart a pacifist, but he was also the biggest kid in the class and very strong. He could literally pick up anyone, carry him for a while, and then let him down as if he had been carrying a small stick of dry wood.

In my own way, I was in charge of the classroom. Any student who challenged me would soon cause me to be led to the headmaster's office. My father never understood that my classroom pranks, and even my fights, were simply my attempt to create a more entertaining atmosphere in what I found to be a very dull place.

Omar, Hossam, Ibrahim, and I liked to pick on the trainees, mostly recent university graduates wanting to become teachers. They had been sent to us to train them, and so we did. We pulled so many pranks. It would never take long to be called to the headmaster's office. My father had the teacher punish me twice as hard as the other students so no one could say that the son of the headmaster was getting special treatment. He made it very clear during those times that he was not impressed with me.

When we had tests, Omar and Hossam—and sometimes Ibrahim, too—usually just wrote down their names and waited until I had finished answering the questions. We would watch for the teacher to turn, then exchange papers so they could see what I had done. Ibrahim did not need me as much as Hossam and Omar did.

I knew this was wrong, and if my father found out he would kill me, but that was what made it fun. It gave me a kind of thrill. It made me feel like all those grown-ups I saw around me—though definitely not my father—who lied, embezzled, paid bribes, and demanded favors. I wanted to understand that immorality by being part of it.

And yet, my real focus in school was on learning the next new popular song or getting the next new hairstyle. Also, I set out to organize the school news program and be its newsreader, to play volleyball, football, basketball, and to attempt to be a part of every team we had. I even joined the music team and represented my school in a singing competition and won the top prize. This did not impress my father.

* * *

The girls at school never talked to the boys unless their families had some connection. The boys, of course, tried to flirt with the girls as they went to their classes, until the teachers came and chased them away. The girls and boys never got to hang out together. We were always sealed off. I disliked that, because I wanted to talk to some of them.

Without a real girl to talk to, I had to rely on my imagination. I vowed to myself that one day I would find a woman to fall in love with. For now, I would just practice writing love letters. I enjoyed writing. The Arabic language has amazing tools for describing things in the most pure and beautiful of ways. I was deeply moved by the poems that had been written during the Andalusian era. I began writing my own poems and love letters, even though I had no one to send them to. I hid them in my desk drawer where I kept my money and my football albums, and locked it so my siblings could not see them.

I was living in a world of my own, drawing my own dreams, waiting for the time when I could go to university abroad. My second-oldest brother, Yazan, was then preparing to join my oldest brother in Germany. I would have to wait for my sister to go before it was my turn.

When we reached seventh grade, the girls moved to their own school, and my classmates and I started to have mustaches. It was weird, and we could not do anything about it. It was in biology class that I learned I was now a teenager. Whenever the teacher talked about private parts, the entire class would go into a state of deep silence. It was funny to watch how everyone tried to act as if they had no idea what the teacher was talking about.

A team of my classmates and I once competed against a group from the girls' school, some of whom had been our classmates in sixth grade. We had to answer questions about all the subjects we were studying. If a team gave a wrong answer—usually the boys— the other team could answer it for them and get more points. The girls beat us miserably. We were crushed. I was trying hard, but neither my teammates nor I could compete.

"I should have brought Omar and Hossam. They would have known more than you," I said to my teammates who were supposed to be the smart ones in my class. I was very embarrassed when we lost, especially when my father began gently mocking us. He had just found something new to mention at home when my thoughts were focused on football players instead of studying.

* * *

When I had a free class at the end of the day, I volunteered to go buy my grandmother's medicine. My father would excuse me from school early, and off I would go to the clinic. My goal was to get the medicine as fast as possible and still have time to play a video game. I would even cut in line or befriend the reception staff in order to get what I had come for and leave quickly.

A café near the clinic had big video games which ran with special tokens that were sold at the entrance. All the kids who could not afford to play or had run out of tokens stood by and advised the others on how to play. That broke my concentration and I would end up losing. Since I could not leave as a loser, I had to play again. And again. And again, until I at last managed to move the stick just right to pull off a *hadouken* move, like in the *Street Fighter* video game. Before I knew it, it would be dark already.

As I sped home, I started preparing my explanations. I knew that one of my parents was going to be waiting for me at the gate, ready to yell at me. I only hoped that none of my siblings would be present. I soon found myself praying for some major event to occur to distract my father.

For all the stresses between us, I was my father's man. I managed his e-mails and wired money when he needed to have some sent while he was busy at school. I liked that he trusted me with a lot of money in my pocket. He sent me to Gaza City on my own, which made me feel very mature. Sometimes I held the phone while he was talking on it. Sometimes I took notes while he spoke. He had a mobile phone, but he told me to carry it for him. I always kept it hanging out of my trouser pocket so that people could see it. That was the cool thing to do.

"What is that on your waist?" he teasingly asked me after one of his calls. "An air conditioner?"

"No," I said, easily riled. "It's your mobile telephone." He nodded his head as if he were happy I knew the difference.

At some point, he decided he needed an Israeli mobile phone, too. Sometimes it was hard to make calls into Israel with a Palestinian phone. I suddenly had two phones on my waist. Of course, he had to mention, "Oh, I see you now have two air conditioners." He knew, though, that I still liked being his soldier on the field.

Part Two

INFERNO

"You shall not accept any information, unless you verify
it for yourself. I have given you the hearing, the eyesight,
and the brain, and you are responsible for using them."
—The Holy Quran, Surat Al-Isra (17:36)

4

Intifada

One breezy morning in early September 2000, I was standing in front of the school talking to Omar. Our first class was starting in a couple of minutes and as usual he had not done his homework. We were brainstorming about what he should do in the short time he had before the teacher arrived. Suddenly, older students from a neighboring high school were at the gate to our school, shouting "Allahu Akbar, Allahu Akbar"—"God is great, God is great."

Was it a religious day that I did not know about? Everyone was shouting that the peace process had collapsed. I had never heard the term "peace process," so had no clue what they were talking about. "Peace process?" I wondered. "The one between me and my father?"

Then, like a dam bursting, a flood of students from my school started pouring out the main door and heading for the street like they all had some place in mind where they wanted to go.

I looked at Omar with question marks in my eyes.

"It's Intifada," Omar said very calmly, like he was an expert.

"So?"

"So, we will get many days off from school," he revealed jubilantly as he let the crowd sweep him out the gate and into the street.

I was very happy about having my first no-school day and started to go after him, but instead I decided to go see my father. Wrong move.

"Classes are suspended, Yousef," he said as I walked into his office. "Don't use the Intifada as an excuse to not learn anything today. Go home and study." He, of course, was staying at the school.

I pretended I did not hear him, and ran out of his office looking for Omar and the boys to enjoy our day off. Omar, who never liked getting out of bed in the morning anyway, said he was going home. Some of my other classmates were heading to the base near my house to throw stones at the soldiers until they got bored and did something else. I knew not to do that. For one thing, my father would be very angry with me. For another, none of the students who were planning to throw the stones lived next to the soldiers like I did.

My mother had told me about the first Intifada, the Palestinian uprising against the occupation that had started two years before I was born.

"Intifada" literally means "a shaking," but it had come to mean an uprising by the Palestinian people against the Israeli occupation. That first Intifada had started in Gaza and quickly swept through the West Bank. I guess the young people saw that they had to fight and rebel against those who had taken their land. Hundreds of young men went out on the streets throwing stones at the soldiers.

Now there was a second Intifada. As I was hustling through the crowd of older students, a large bullet-shaped object, about the size of a potato, skidded along the road right in front of me.

"Throw it back now, Allahu Akbar!" someone shouted. Here was my chance to be the big man. I picked it up to toss it. But, oh, it was like I had picked up a boiling potato. I tried to throw it back toward the soldiers, but it was too hot for me to keep in my hands. A thick white smoke began bursting from it. The smoke stormed my lungs and nose. My eyes were flooding with tears, and my whole face was covered with something yellow and felt like it was on fire. Quickly, I passed the object to an older boy who was wearing thick gloves.

I ran through the crowd and back all the way to the house. I did my best to be invisible as I locked myself in the bathroom and splashed water all over my face like a madman. All I did was get the front of my shirt soaked while my eyes still stung. I whipped off my shirt and tried to put my face under the tap, which just made matters worse. I was glad that no one had seen me. It was clear that I was not doing my ablutions.

Finally, the burning fire in my eyes began to calm down. I wiped my face on the part of the shirt that was still dry and put it back on. I snuck out of the bathroom carefully, looking in every direction like a commando to make sure that neither my grandmother nor anyone else saw me. I decided to go to the café and play some more video games. It was a day off after all.

That evening, as everyone was home early, we ate early and then watched the news. Over and over again, they showed a clip of the Israeli prime minister, Ariel Sharon, entering Haram al-Sharif in Jerusalem, the site of the Al-Aqsa Mosque, one of the three holiest places for Muslims. Although Jews were allowed to visit Haram al-Sharif, they were forbidden from praying there. Actually, even most of the Palestinians could not go pray there without obtaining a security permit from the Israeli army. Sharon's visit was seen as provocative. The "peace process" had immediately collapsed.

My older sister started asking a lot of questions that began with "Why." Nobody answered.

My grandmother Zana had lived through the British Mandate, and the wars in 1948, 1967 and 1973, as well as the first Intifada in 1987 and now this second one. To her it was all just that same old boring movie being played yet again. She waved her hand at my sister's questions and refused to talk about it.

My mother, though, started talking about her brother Mujeeb again.

"Your uncle served seven years in prison," my mother said with a fresh sense of disbelief on her face as if it had just been last week. "For throwing stones! Don't do these things, Yousef. You better promise." I could not understand why she was looking at me when I had two older brothers in the room. It was like I was the reason Uncle Mujeeb had been arrested. I had not even been born yet. But I had heard the story told the same way so many times over so many years that I felt like I really was there when it happened. I was never sure if they told it as a warning to us younger kids or as a cry of pain from the adults who were powerless to stop a terrible injustice.

Uncle Mujeeb was my mother's younger brother—a laid-back, blond-haired, good-looking guy. He always had a cheerful smile and dressed well. He and his youngest brother spent a lot of time at my house as teens because their father had died when they were young. They enjoyed being with my parents. My father helped fill some of the space left by their father. My mother was like a second mother to them.

My uncle was barely sixteen years old when they came for him. Uncle Mujeeb told me many years later that on the night he was arrested, he was watching World Championship Wrestling, trying to decide whether the fights were real or fake. As soon as he heard the soldiers at the door, he bolted from the room and hid in the garden. They had already arrested his older brother, Uncle Saed, to force Mujeeb to turn himself in.

My father attempted to talk the soldiers out of arresting him by saying that he was a good kid and well behaved; they should give him a break. But the soldiers were unrelenting and the Israeli intelligence officer with them said that they would keep his older brother locked up until Mujeeb turned himself in. When they finally left without him, Mujeeb came out of hiding and stood before my father. He had been crouching under some pomegranate trees. The soldiers would not go into trees like that where they could not provide themselves enough cover.

He sat at the table in the kitchen while my mother fed him. My father sat with him and told him quietly that he really had no options.

"Maybe if you turn yourself in, they will take that as a good sign and let you go free along with your brother." Mujeeb just ate the lamb stew that my mother knew he loved, and said nothing.

"I am worried," my father went on, "that if you keep on being chased, the soldiers might use that as an excuse to shoot you." My mother was standing by the sink and took in a deep breath.

"You might go to prison, but at least you would be alive. In prison, we might be able to do something for you." Left unsaid was that if he were dead, they could not.

Mujeeb listened to my father, then sat for a while in silence. As he got up, he told them, "I'm going to take a shower. It might be a while before I get another one." My father nodded.

After Mujeeb had bathed, he put on layers and layers of clothes, because he knew he would be beaten. Then he put on his fancy jacket, which had the American flag on its back. He never wore it, so my mother had used it to hide her gold jewelry in its pockets. While my father was in the car waiting to take him to the military base in downtown Deir al-Balah, and my mother was assuring him she would come see him in prison, she suddenly stopped.

"Oh wait," she said, "Let me check that inside pocket. I might have left some jewelry in there!"

I always laughed when I heard that part of the story. It must have been funny to hear her say that after all the kissing and hugging. Well, not very funny at the time, but afterward maybe.

Before Mujeeb got into the car, my father searched him. He had a knife hidden in his trousers. My father took it from him without saying anything, then buried it at the side of the driveway until he came back.

"They will use everything you give them to make it worse," he said as he came back to the car. "Don't let them."

They drove to the base. Mujeeb turned himself in. His older brother was released soon after, and returned home safely. Uncle Mujeeb never had a trial, but was found guilty of throwing stones and sentenced to fifteen years in prison.

My mother said that the Israeli soldiers would sometimes dress in jeans and T-shirts and pretend to be throwing stones, but in reality, they were gathering information: names, where the Palestinians lived, and what the plans were for the next day's stone-throwing.

"That's how they caught most of them," she said.

Mujeeb was set free after seven years when the first Intifada ended in 1993. I was excited. I was five years old and suddenly I had a new uncle I had never met. My father brought him home from the prison to a large party at our house. My mother hugged him for a long time at the door of the house. And his mother, Sitie, who almost never cried, could not control herself when she saw him. He was considered a hero by a lot of people because he had fought back against the occupiers. Maybe. But his life was never the same afterward. Whenever I heard his story in the years after, I tried to imagine what it was like to be sixteen and in prison, to open your eyes every morning and face another day behind bars. I could hardly believe that a teenager had been put in a real prison for throwing stones.

Around the same time, Yasser Arafat signed a peace deal that created what are now known as "The Palestinian Territories." He sent a message to his people that every household in Palestine was to make peace with every household in Israel. This was a very new idea.

"Coexistence is the only way we will ever have peace," my father said to everybody who came to the house. "They are here. We are here. No one is leaving. Coexistence has to be the new norm."

I began to hear my father talking about an organization that I think was called the Sons of Abraham. Its members were Christians, Jews and Muslims. He went to their events in Jerusalem and Ramallah a couple of times, and always came back from those places with large bags of sweets. So, I thought these meetings were a very good thing. Once, he brought home a framed verse from one of their conferences that he placed in front of the mirror on my mother's coiffure table. It said:

Peace begins at Home
Peace begins in Me
Peace begins in You
Peace begins in Her
Peace begins in Him
Peace begins in Them

Not everyone agreed with my father, of course. To the outside world, the Palestinians might have appeared to be of a single mind. Every Palestinian, however, had his own reaction to the Israeli occupation, depending on his own circumstances and experience. Some chose violence, but my father believed that violence was self-defeating.

"Violence only leads to more violence," was his mantra.

Palestinians who disagreed with him never challenged him publicly because he was held in such high respect in the community. He was a passionate Muslim, and everybody knew that. They could see the deep pleasure he took in his prayers and the sense of serenity that his close relationship to God brought him. The secret of his strength was his relationship to God as a Muslim.

Several of his closest friends strongly opposed his point of view, which he knew. My father accepted all people, whatever their political beliefs. One of the teachers in our school, my history teacher, was committed to Hamas. He wore the traditional ankle-length white *jalabiyyah* beneath a western-styled blazer and a plain white *keffiyeh* on his head. He was a sheikh and resisted singing the Palestinian national anthem because he saw it as an act of capitulation to how things were. Yet his respect for my father knew no bounds. They never argued. Mostly they told jokes when they were together.

There was an English teacher who was also very close to my father and helped him run the school. He wore a beard and profoundly disagreed with some of my father's ideas, though he did not affiliate with Hamas. Both of them were highly educated and sophisticated, however, and loved to talk about the books they had read.

The two teachers always came to our house on both Eid days to pay their respects to my father. My father always returned the courtesy by visiting their homes as well. Politics never came between them.

Both of the teachers were refugees whose families had fled other parts of Palestine after the formation of Israel in 1948. Unlike my family, they had experienced the special bitterness of being refugees. My father understood that. They all knew, though, that there was no reason for them to argue among themselves, that the cause of their problems was not with each other. They practiced their own form of coexistence.

* * *

One morning the house started filling early. Every woman who was any kind of relative was in the kitchen with my mother. Many had brought food; others were working over the stove. It was 1998 and I was nine years old, so it was my job to help carry dishes and trays up to the roof where my father and my older brothers were with the husbands of the women in the kitchen, along with several of the teachers from my father's school and some of the neighbors.

Hanging on the wall on the side of the house facing the Salah al-Din Highway were Palestinian and American flags. In fact, the whole length of the highway was lined with flags. I wondered where so many American flags had come from. Did people keep them in their houses and not tell anybody?

Somebody had carried our television up from our living room. A sound system was playing Arab pop music. No one paid much attention to either. Everybody was too busy talking and laughing. It felt like a wedding party.

Then somebody yelled, "They're coming. They're coming!" I raced down the three flights to the kitchen to tell my mother and the other women. They stopped everything they were doing, and ran up the stairs, except for my father's mother who just waved them on up and went out on the terrace.

We had never had so many people on our roof before, all trying to get close to the highway side to see the motorcade. First came what looked like a whole army of motorcycles. Even from a distance I could hear their roar. It was like having a jet plane driving down the road. They made an arrow formation as they swept south along the highway. Behind them were several black vehicles that looked like jeeps, but were much classier. They all had flashing lights.

Then came what we were all there to see: President Bill Clinton's car. Big and heavy and looking impregnable, it had two flags stationed above its headlights that flew stiff out in the wind as the limo sped toward us.

All the grown-ups on the roof started cheering and screaming like they were little kids. Some waved small flags—it did not seem to matter from which country, they just whipped them back and forth. This was the first time a president of the United States of America had ever visited Gaza. Indeed, Clinton was the first US president to acknowledge the suffering of the Palestinians at the hands of the Israelis. Now he was driving the length of the Strip with Yasser Arafat. People were filled with hope and optimism about the future, and they were making noise!

Clinton's limousine did not slow down just because we were all on the roof waving flags. And the truth was, it was impossible to see who was actually in it. By the time it reached us, the motorcycles and all the noise they were making were already far down the road.

Then they were gone. They were headed to Rafah at the southern tip of Gaza where Clinton and Arafat formally opened the new airport built with the help of the international community. Having our own airport meant that we would no longer have to go all the way to Jordan or Egypt to get on a plane because Palestinians were rarely allowed to fly out of Tel Aviv's Ben Gurion Airport.

The men on the roof began hugging each other; some made little dancing steps. The women began passing the food around. When the motorcade reached Rafah, a lot of guests crowded around the TV to watch what was happening there, but it was hard to see the screen in the bright daylight on the roof.

As I sat with one of my cousins eating some of my mother's good food, I began to think about the United States and the role it played in my country. It did not take long for me to realize that the US was a third player in the political triangle that affected my people.

A few months later we heard that Clinton was in trouble for having sex with a Jewish woman at the White House. The word on our streets was that the Jews, or, to be specific, the Israelis, were punishing him for having come to support our cause.

* * *

By the time the second Intifada started less than two years later, all the smiles of that day had faded. The soldiers on the base next to us no longer partied as they usually had on Saturdays. To me, they looked scarier than before. Their weapons gave me the feeling they were almost alive and looking for action. I could just sense that it would not be long before I saw them being fired.

These feelings were intensified when my uncle Yousef, after whom I was named and who had cut down our orange grove, decided to leave the house he had built. I could not believe that after all the trouble he had caused between himself and my father, he was just going to go. Like that.

I had grown up with the story about how Uncle Yousef had returned home from Libya after the death of my grandfather years before, hoping that he and my father would split the land equally. However, my grandfather had registered all the land in my father's name, so my father had the legal right to decide how it would be distributed. Apparently, my grandfather did not trust Yousef to treat all his siblings justly.

Yousef claimed that because his two sisters were married, they should not get any of the land.

"Their husbands will just end up with it anyway," he argued. "So, what's the good of giving it to them?" No matter how hard my father tried to persuade him otherwise, Yousef insisted that the sisters be denied any share of the land. My father refused to do that.

A couple of years before the second Intifada started, Yousef grew so furious that he recruited some tough guys who were distant cousins to beat up my father in a last-ditch effort to bring him around. Even then, my father did not give in.

My mother told me in a whisper, "They beat him badly, but he still insisted that everyone should get their fair share." When my mother's family wanted to intervene and physically defend my father, he refused their help. This story always filled me with deep respect for my father, though I could not help but wonder why it is the good person who gets hit all the time.

The day Uncle Yousef came to announce that he was taking his family and leaving, he and my father sat in the garden surrounded by greenery. My uncle was a television news anchor. He was always sharply dressed to make himself look like a celebrity, unlike my father—his kid brother, Khalil—who never had to lift a finger to look very cool.

As a gentle breeze ruffled the palms overhead, the cups of hot mint tea that had been set on the low table in front of them went untouched. It was clear this was a final conversation. Yousef was being rude toward my father.

"Where will you go?" my father asked when Yousef seemed to have run out of words.

"With my in-laws in Deir," Yousef spat arrogantly.

My father chuckled sadly. In Gaza, extended families live together—grown siblings and their children, multiple generations adding on and laying down roots beside one another. Despite all the problems Yousef had caused, my father did not want him to leave, even to move only a few miles away. Within days, Yousef loaded up a truck with all of his belongings, his wife, and his children, and drove off.

I did not mind seeing him go. In fact, it had been hard to grow up knowing that I was named after him. My father had thought the world of his brother Yousef when he chose my name, only to be disappointed years later. Whenever I was told off for doing something wrong or failing to do something right, my siblings would tease me about being named after a bad man. This, of course, gave me yet another reason for always insisting on doing things my way. I was determined to liberate myself from Uncle Yousef's shame and restore my name's reputation in the house. Uncle Yousef had never spoken to me or even done as little as shake my hand or give me a hug. Now he was abandoning his house and his land.

* * *

Even with an Intifada, life still went on. We still had to go to school. But indolent students like me secretly hoped that the guys in the upper classes would break through the gate again and shout, "Allahu Akbar." Just one loud "Allahu Akbar," and it was free day. I was like the students in other countries who pray for a hurricane or a heavy snowstorm to cause a day off. When we heard an "Allahu Akbar," we would all run outside, no matter how hard the teachers tried to stop us.

5

The Soldiers

My mother was the first to realize things were different. From her window in the kitchen she could see that, as the Intifada wore on, the number of vehicles going on and off the base was steadily increasing. Every few days, we could see more soldiers in the tower, on the tanks, or just walking around.

"Khalil, what is going on over there?" she asked my father, as if he were the base commander.

My father was puzzled, because this was not the first time that peace efforts had failed, so he saw no reason for the soldiers to be acting as if they were at war.

Sometimes I went up on the third floor of our house to watch them through the unfinished windows. I told myself that they were the same soldiers who danced and partied all the time and upset my mother with their loud music and teenage antics.

Now, as I looked at the soldiers jumping off their tanks when they arrived at the base, I tried to assure myself they would never do anything bad to us. We had been living next to one another for eight years. Why would they ever hurt us? We had never done anything to them. I told myself, "They'd never shoot at us or try to kill us." It seemed impossible, but I could not help wondering. I was eleven years old and found myself concerned that I could no longer take my beautiful world for granted.

As the weeks passed, I became increasingly fearful. For the first time in my life, I was afraid of the dark. It was like I was expecting a demon to appear right from hell. The elders in my family were always talking about how much better the old days were and how people had cared about each other. All that talk made me feel that something awful was about to happen.

I tried to distract myself with happier thoughts. I nagged my mother to buy me a new bicycle. I told her that a bike would let me do her errands faster, that I could get home from school faster, that I would give my little brothers rides. It took me forever to convince her, but in the end, I got my wish. I was so happy to have it.

I take good care of my possessions, and I took good care of that bike. I usually parked it in the garage next to my father's Opel to be sure it was safe, but one night I had forgotten to do so. My father was out, and I was with my mother and grandmother in the kitchen when we suddenly heard guns firing outside our house. Lots of them, all at once.

The shots paralyzed my mind. The sharpness of their sound gave my head no time to process a single thought. I ran to the window to look and find out what was happening. I saw the soldier on the back watchtower behind our kitchen. He was shooting and shooting.

"Get away from the window, Yousef!" my grandmother shouted.

I dropped to the floor. Swarms of bullets started clipping through the kitchen window where I had just been standing.

"Everyone down!" my mother yelled to my three brothers and three sisters in the other rooms. "On the floor! *Yalla, yalla* [fast, fast]!" My siblings, my grandmother Zana, and Uncle Mujeeb, who was visiting us, all flung themselves down.

Bullets were flying like mad bats over our heads. I could see them as they crashed into the walls and doors. Bits of plaster and splinters of wood exploded from where they hit. My little sister Zana was crying, and my mother was calling our names to be sure we were alive. Soon we were all crawling on the floor, struggling to look calm and to get to the living room, where we would be out of the direct line of fire. After living for so long with a looming sense of threat, this was action time for real.

Then I remembered my bike. I had left it in the driveway. I got up to sprint outside to get it.

"My bicycle, my bicycle!" I shouted, as I headed toward the door. Uncle Mujeeb caught the edge of my shirt and grabbed me.

"Stay down! Forget the bicycle!" he thundered.

"OK, so you will buy me a new one?" I demanded.

He said nothing, only pushed me to the floor and kept a heavy hand fixed on my shoulder for several minutes as he cowered behind the TV.

I gave up. Bullets seemed like they were everywhere. I could smell gunpowder. Lots of it. For the first time in my life, I was truly scared, and I knew everyone else was as well, even if they pretended otherwise.

Much as he tried to hide it, Uncle Mujeeb was the most frightened of us all. From where he was pinning me down, I could see how scared he looked. Part of me wanted to laugh. He was always telling stories from his time in prison about how tough he was, and yet it was clear to me that he was as terrified as I was.

We stayed like that for what felt like an eternity until, finally, the shooting stopped.

A short time later my father walked into the house, strolling in casually as if he had just come home from a meeting at school. In spite of the lingering odor of the gun smoke, we could smell the familiar scent of his Azzaro cologne, and a strong sense of safety filled the house once again. As he picked up Zana, he tried to calm us: "Everything is going to be OK. Don't be afraid."

Soon he was saying that to us all the time. *Don't be afraid, don't be afraid. It will be all right.* Those words never made any sense to me, for I could see that what he was saying was not true. For the first time in my life, I had to question whether my father really knew what he was talking about.

That was when my mother's family started insisting that we all come to live with them. They were worried about our safety on what had become the front line of an unexpected war. Her mother, Grandmother Sitie, has a forceful personality. She tried every approach.

First, directly: "Khalil, it just is not safe for you and the children as long as those soldiers are there. Why not come into Deir for a while until things quiet down?"

Then, with more persuasion: "Khalil, I have so many rooms in this house and most of them are empty. You would be doing me a favor to live here. I would welcome the company."

Then more subtly: "Khalil, how can your children study with all that noise? They have their exams coming up. They need some peace if they want to score high marks."

Nothing worked. My father even had a small house in downtown Deir to which he could have moved us if he had wanted to.

"Leave?" he would answer with a light laugh. "There is no reason to leave. This is our home."

My mother had her fears, but she always stood behind my father. For her, he was everything. He made it very clear that he was not going anywhere. I knew she would never abandon him, despite the dangers of staying in our house.

* * *

The shooting soon became an everyday routine and our windows kept getting shot out. We started replacing the glass with plastic sheeting, which was a lot cheaper. That really annoyed me. For years, my parents had forbidden me to play football near the veranda windows for fear of breaking the glass. Now the windows were all shattered anyway and the Israeli bullets kept ripping through the plastic again and again. Every time that happened, my father made us stretch new plastic over the window frames. When it was my turn, I did it very slowly and very grudgingly. I hated doing chores anyway, and I especially disliked this one.

It had become very dangerous to be in our neighborhood, even for those who lived farther from the base than we did. Tanks were frequently driving in and out, and Apache helicopters swarmed in the sky above our land. The Palestinians were busy fighting the occupation and the Israeli soldiers were busy fighting the Palestinians.

Every day hundreds of Palestinians were out on the streets demonstrating, and every day I would hear stories of many of them being shot and killed. Posters of young men were being pasted on the walls of buildings all over Gaza. They were called martyrs.

* * *

Whenever the shooting began again, my father would turn on the radio to listen to the news while sipping his black tea. We would all gather around him, while he assured us with his usual confident voice, "It will stop, it will stop."

One morning after a night of endless firing, I went to the kitchen where my mother was preparing breakfast. We were having our usual breakfast kind of talk.

"If you want eggs, you will have to go out and get some. Bring some for the twins, too."

"I don't want eggs today," I replied, because I hated going to the chicken house first thing in the morning and having to breathe its sour smell before I was even awake. Maybe after I had had some tea.

As I opened the window shutters, I saw the body of a young Palestinian, a teenager not much older than me, on the fence the soldiers had built around their base.

I just stood there and looked. My mother sensed something was wrong and came and stood next to me.

This was not the first time I had seen the horrific carnage of war, but the way he was strung across the barbed wire like a badly broken puppet, with his arms and legs all twisted in different directions, made me freeze. His AK-47 hung heavy around his neck.

"Why? Why?" my mother said, almost in tears. This was a question I heard all the time. There was never an answer.

As I zoomed in on the body, I saw an Israeli robot roll up and jerk the young man's body off the fence. Then the robot dragged him over the rough ground and across the field to where a Palestinian ambulance was waiting to collect his lifeless body. Afterward the robot came back to get the body of a second young man whom I had not seen at first, and dragged him in the same way across the field to the ambulance.

Grandmother Zana had come into the kitchen and was standing next to me, watching, but saying nothing. I was filled with disgust as I saw the bodies being hauled through the dust. Was there no respect for the dead? They were just teenagers. They might have been crazy for thinking they would succeed in infiltrating the base, but they were not bad people. They were targeting military officers who were occupying their land. They were in a war. Worst of all, they were described as terrorists to the rest of the world. The Israeli soldiers had never looked so shameful in my eyes.

I had lost all interest in eating, but I did go to get some eggs for my younger siblings. As I walked out to the chicken coop, which was just feet away from one of the watchtowers, I thought of Muhammad al-Durrah, the twelve-year-old boy who had been killed by the Israelis as he and his father hid behind a piece of concrete on a pavement in Gaza. They had been trapped by Israeli bullets on the second day of the current Intifada.

A Palestinian cameraman who was working for French television had filmed Muhammad's father waving his pack of cigarettes and begging the shooters to stop while his son squeezed up against him, trying to stay safe from the bullets. Then, after a minute, while we watched, Muhammad lay dead across his father's legs. It was the first time I had seen a child shot for no reason. In fact, I had never actually seen anyone being shot before. It made a deep and lasting impression on me.

By now, there was no way to predict when there would be shooting outside the house. It could happen in the morning or at night. Our house blocked the base from the rest of the neighborhood. The soldiers started pounding our house day after day. Maybe they were hoping we would leave. The walls shook as the house was hit with waves of bullets and small missiles. This, I thought, was the truth standing between me and the words of my father.

It only got worse, until the soldiers realized we were not going to leave our home.

One day, about three months after the Intifada had begun, we heard loud knocking on the door. A voice with a European accent demanded, "Open za door, open za door."

They banged so hard that the door almost fell in. As the soldiers poured across the threshold, one of them asked, "Where is Khalil? Get Khalil." They had come to the house other times to complain about some problem here and there, but this time was different. This time they had painted their faces in black and green and looked as if they were out to invade Vietnam.

When my father appeared, he extended his hand and greeted them warmly, as he might have welcomed our Palestinian neighbors.

"Hello, welcome to my house," my father said.

They completely ignored his gesture. Instead, the soldier in charge kept his hands tightly gripped around his M-16.

"You and your family have to leave the house." His hair was cut short and his eyes were heavy-lidded. He had the look of somebody who would rather be somewhere else.

"Is there a problem?" my father asked.

"This house is not safe for your children, Khalil." He said it as if our safety were the only thing he cared about.

"Thank you for your consideration," my father responded in his most kindly manner, "but of course it is safe. This is their home."

"This is not possible." The commanding officer closed his heavy-lidded eyes for a moment. "Get your documents and any personal valuables and leave," he insisted.

"In fact, *that* is not possible, because we live here and we are not going anywhere," my father said, like he was giving directions to a lost driver.

"Khalil, be reasonable. You see what is going on here. Other people are leaving, and for good reason."

"Are you telling me that the settlers are moving away?" My father's face was alight with a broad smile.

"Khalil, I have not come here to tell jokes. You have to leave."

"Thank you, but I don't care where anyone else goes. This is where we are from, and this is where we live," he repeated. "Perhaps you and your men would like some tea."

My father was determined not to abandon our land. He did not want our family to become refugees. For my father, becoming a refugee was worse than death. He did not want his children clinging to old deeds and the house keys of confiscated properties with hopes they would be returned one day.

The soldiers left us that day with agitation on their backs; we knew they would be back. I kept wondering what was going to happen to the life I had known—the farm, football, my stickers, and all the other things I loved. What would become of the heaven in which I had lived? Trees were being cut down wherever the soldiers needed to build a new road for their tanks and bulldozers. I could go upstairs and watch nearby houses being demolished. I watched Uncle Yousef's house get demolished and thought that soon it would be our turn.

A few days later, without any explanation, our "neighbors" from the base became our "house guests." They simply walked into our home as if we were not there, moved into the upper two floors, and set up their guns on the rooftop. They had taken over our house in the blink of an eye, just like that.

My older sister shouted at them, "Where are you going?" She tried to block their way, but they just walked past without even acknowledging her presence. Right then and there, they took over our house. From our top floor they could see the whole neighborhood.

They smashed holes through the upstairs walls to set up gun positions. They covered all the windows with camouflage netting and installed automatic machine guns at each corner of the roof. The guns had cameras attached to them so they could shoot whenever they registered "danger." We assumed that the guns could shoot on their own, though we were not sure of that. Instinctively, I knew never to be in their sights.

When the soldiers told my father that they were going to stay in the house, his response was, "For how long will you be our guests?"

A soldier responded, without hesitation, "Until this is over." With that, he retreated upstairs and shut the door.

"That is a good sign," my father declared. "*Until it is over* means that it will be over some day, and they know it."

It may have been a good sign to my father, but to the rest of us it was a disaster. And it would remain a disaster, no matter how hard he tried to make it seem otherwise. I felt sorry for my grandmother having to watch her son deal with the soldiers every day. But, like my father, she refused to leave. Though she was frail, she, too, had a will of steel.

For the first few nights, I was intrigued to watch the soldiers and their guns at close range. As time went on, though, and they became increasingly aggressive toward us, I was disgusted to think I had ever considered them interesting.

The soldiers now officially occupied our house, and we were now their prisoners. They acted as if they were in the most dangerous house on earth and were furious if we so much as sneezed or moved suddenly. Sometimes I would sneeze just to provoke a reaction, but I stopped when I realized my father had noticed what I was doing.

Sometimes I would fantasize about snatching an M-16 from one of the soldiers and shooting all of them. I quickly discarded that thought, however. If I had tried something like that and survived, my father would have punished me so much. Maybe it was perfectly acceptable to imagine teaching the soldiers a lesson by using their own weapons, but it went against everything my father believed.

We were allowed to leave the house only during the day, and then only to go to school. I could no longer go out on the land to find a quiet place to think, and I could no longer climb up the olive tree because it made them "nervous." The area behind the house was totally off limits to us, as was the upstairs. We were told we would be shot if we ventured there. The worst thing was that we could no longer have any guests. Our house had always been filled with guests.

Some came for a cup of tea, some stayed for a week. This was normal for us. Our guests were not only Palestinian relatives and friends, but also German, French, and British students who would come stay at our house while doing their exchange studies in Gaza.

One night soon after the soldiers had taken over our house, they came downstairs and herded all ten of us into one of the two living rooms, and told us that we were going to sleep there for the night. There were six walls in our house between that room and the base where most of the shooting was coming from, so arguably it was the safest place for us. We presumed there was going to be some special operation for that night.

One of them crouched into a firing position with his M-16 pointed at us as we moved some couches around. From the veranda we gathered some of the *freash* on which we sometimes slept. Nobody said anything except my mother who told each of us to pick a spot in the room. I chose the corner just right in front of the door so I could watch the soldiers going up and down when they left the door open.

For one night, this was kind of cool, even if the room was small for all ten of us. The next night, though, they made us do the same again. And the next night. It became our way of living.

We never knew, though, when the soldiers were going to lock us in. Sometimes it could be as early as 3:00 PM. Sometimes 8:00 PM or as late as 2:00 AM. We never knew. If they had not yet come when I wanted to sleep, I went to my bedroom. Maybe one night in five I would be able to sleep the whole night there. Other nights we went to sleep in our own beds, only to have the soldiers wake us all up in the middle of the night and herd us into the living room, shouting at us to move fast.

If we heard *tak, tak, tak*, we knew some Palestinian was shooting at the base, and the soldiers would soon be locking us in, and we would huddle together and listen to the cacophony of war outside.

Sometimes when the soldiers came, I could hear my grandmother on the veranda, praying in a soft voice. My father would ask us all to hurry and move quickly to the living room. They gave us about fifteen minutes at most. Then, as soon as we settled into our places there, all we heard was shooting and shooting and shooting. My siblings and I just went back to sleep on whatever spot we had claimed, but my parents and grandmother would usually stay up through the night wondering what might be happening outside. Sometimes, my grandmother might suddenly start asking if we had closed the animal shed or if we had fed the animals. All we could do was pretend we did not hear her. Sometimes we really could not hear her, because the shooting was so loud.

This became our nightly routine. I tried to predict when they would lock us in the living room by carefully watching the news of what was going on around Deir and the rest of Gaza. Also, I secretly observed the base from between the window shades to predict whether they would come that night or not. They came so many nights, I was almost always right.

I could never predict, though, when they would let us out in the morning. Some days they did not come at all. Sometimes they would keep us locked up in the living room for a week or two and not let us go to school. Other times, we would get to school late. My father hated that. He did not like setting a bad example for the rest of the teachers, whom he expected to be on time.

"Excuse me," he said from the living room doorway to a soldier who appeared to be in command. "As you know, I am the headmaster of a high school where the classes are about to start. I need to call them and explain that I will be late today."

When the soldiers locked us in the living room, they also took away all our phones—even the two landline phones—and threw them in a corner by the front door. On top of all the other stresses, this made us feel dangerously isolated.

"You can call when we leave," the soldier told him.

This happened so often, it became a routine. Many times, my father had to speak to the soldiers through the locked door, asking to be let out or at least to use the phone. Often they did not even respond.

One of the strangest parts of their occupation was that we were not always sure whether the soldiers were upstairs or not. Usually we could hear them, but not always. For the most part, they used our stairs to come and go, but at other times they used a folding ladder at the back of the house. It was hard to know which was worse: knowing they were there or not being sure. We did not dare go up to check.

Once, when a British journalist was visiting, there was a loud scraping noise from the floor above.

"Ah," my father said, without looking up, "they are rearranging the furniture. They took a table from downstairs while we were asleep. I guess they are making it fit into the room," he joked.

Other times he joked that the soldiers were "working on" the upper floors. With the delays caused by the lawsuits from the settlers, my parents had never been able to finish the construction. The plumbing and wiring had been installed, but never hooked up. There were no windows in the frames. The space was essentially two floors of empty rooms. This had not been a problem at the time, because that space would not be needed until my two oldest brothers were married and brought their wives home to live with us. That was still several years away.

One time the soldiers ordered my father and mother to go with them upstairs and block one of the window openings with cement blocks. The soldiers watched as my parents did what had been ordered, but never said why this was needed.

* * *

When the soldiers were in the house, we could use the kitchen and the bathroom only with their permission. When we went to the bathroom, a soldier stood guard outside. They did not let us close the door when my father and my brothers used it. Some of them at least turned their back to give some "privacy." They let my mother, my grandmother, and my sisters close the door without locking it, but stood there the whole time.

This got old real quick. Even when we needed to go to the bathroom, it was embarrassing to always have to ask, and we would try to hold it in. Sometimes, though, we just had to go.

One night I stood at the locked door of the living room for several minutes trying to get the attention of the soldier on the other side. Maybe he was asleep. Maybe he had stepped away from his post. Maybe he was just being mean-spirited. But he would not let me out. Every time I called to get his attention, I knew I was disturbing the whole family, who were sleeping as best they could. They would start rolling over and pulling up blankets when I called for the soldier.

Finally, I heard the lock turning. I had been shaking my legs for the longest time and sped toward the bathroom.

"Stop," he yelled. I was too afraid not to, but, oh, I had to go so badly. He called upstairs. "One of them has to use the bathroom." Silence from above. I started to count to one hundred slowly, anything to take my mind off things. After a couple of minutes, a soldier came down in full combat gear like I was a one-man hit squad waiting for him.

"*Yalla, ta'al* [come on, let's go]," he said, and I walked as slowly as I could down the hall to the bathroom. With each step, I tried to believe that this night he would leave me alone for a few minutes. Of course, that did not happen.

Sometimes when I needed to go to the bathroom badly, the permission did not always come in time. We had some buckets in the living room for when we had no other option, but it was awful using them, so degrading and humiliating. I tried to squat behind one of the couches when I needed one, but it was so loud and so embarrassing, I just wanted it to end as fast as it could.

When I spent a night in my own bedroom, I liked to sleep wearing only my underwear. For some reason, that made me feel like a grown man. I'm not sure why, exactly; perhaps because I thought a grown man should feel comfortable enough in his own home to be able to undress for bed. So, I really hated the nights when we were forced out of our beds and into the living room to get locked in for the night. A soldier would order my mother to wake me up. I would tell my mother to turn around to give me some privacy while I put on my jeans, but no privacy was ever given.

The soldier would point his very bright torch at me, and my mother would turn on the lights so that I could see what I was doing. But of course, lights were exactly what I did not want at that moment. I just felt so lame, in the way only a teenager can, to have the soldier *and* my mother looking at me in my underwear. Despite the abject humiliation of these moments, there was never any time to discuss this with my mother; she had to get all of us moved into the living room quickly.

Once, when a couple of the soldiers felt that my father had not got us into the living room "on time," one of them smashed his head against the wall. He just acted like nothing had happened, but we were all shocked like it had been our heads that had just got banged against the concrete wall.

There were nights when, after we had left our bedrooms, the soldiers would go and sleep in our beds. One time when my mother was heading to the bathroom during the night, she passed her bedroom and the door was open. "I looked in," she said later, "and there was a soldier in my bed with no clothes on."

The soldiers helped themselves to our belongings without asking. Once I woke up in the night to find my blanket had been snatched from my bed. My father said that that was a good sign.

"If he was cold," he said, "that means he is human."

6

House Guests

I was gradually becoming oblivious to the sounds of the shooting. I began to wonder whether it was healthier to learn to ignore danger or to maintain the ability to fear. It is a terrible thing to live in fear, for it can slowly turn you away from your own humanity and your compassion for others. I was too afraid to apply any compassion. Is one already dead if one no longer fears? So much was going through my mind. I would get scared, not so much for me, but for my father. The soldiers never stopped demeaning him.

One day he came home from school later than we had. He had been holding a teacher's meeting. He drove up the driveway and parked in the garage, just like always. The rest of us were inside doing whatever we were doing.

As he came up the stairs to the house—his house—one of the soldiers stopped him from coming in.

"Take off your clothes," he ordered my father, who laughed. The surreal level of our existence had just hit a new level for him.

"I'm sure you don't mean that," my father said. The soldier pointed his gun at my father and repeated his order.

"Now!" he shouted, as if my father were not standing three feet in front of him. My father looked at him, searching his face for an explanation. There was none.

"May I ask why?" my father queried him quietly, with his eyebrows raised and his palms upturned.

"Because I said so!" the soldier yelled back.

"Then may I step inside the veranda?"

"*Akshav! Akshav!* Right here. Right now. This is the last time I am telling you."

By now the soldier's shouting had brought all of us to the front door. He shouted something into his walkie-talkie in Hebrew.

"*Shu fee* [what is going on]?" my mother asked as my father took off his suit jacket, folded it and set it on the top step beside him.

"Back inside!" the soldier ordered as a couple of other soldiers came racing down the stairs and pushed us back into the living room.

My father unbuttoned his shirt, took it off, folded it, and laid it on top of his jacket. He undid his shoes and slipped his feet out of them.

"My socks?" he asked the soldier.

"Everything!"

He pulled off his socks, then slipped off his trousers and undershirt. All he was wearing was his underwear. He looked at the soldier as if to ask, "These, as well?"

"Turn around." My father turned his back to the soldier. If he were worried about getting shot in the back, he did not show it. He stretched his shoulders and took a deep breath as if he were happy to be getting some sun, though it was late in the afternoon.

"Go inside," the soldier ordered. My father turned around and faced him. And smiled.

"Thank you," he said, then bent to pick up his clothing. As he came through the door, he asked the soldier, "Can I bring you a cup of tea?" The soldier ignored him.

My brothers and sisters and I had all gone inside the house and pretended we had not seen anything. In Palestine, it is very embarrassing to be seen in your underwear, as it is for the people who see you. We all felt a kind of shame for my father. He went straight to his bedroom, got dressed and reappeared a moment later, telling my mother he was ready for lunch, as if nothing had happened.

Once again, we made the mistake of thinking that this was a onetime event. Until it happened again. And again. And again. Sometimes, they hit him if he did not take his clothes off fast enough.

Then my father walked in to join us as if no one had just seen him get hit. He stood in the room completely still, unruffled. Things like that always left me feeling appalled.

* * *

Soon after that, the soldiers decided to search the house on a regular basis after dark. They made my father walk in front of them. With their rifles aimed at his back, they went from room to room, presumably checking for any Palestinian fighters who might be hiding there.

It started when the captain in charge that night demanded, "Open za door! *Kadima, kadima* [Let's go, let's go]!"

"I have to gather my children first before I go upstairs," my father would reply quietly, the tone of his voice making it clear to the captain that it would be better for the soldiers themselves if they allowed him to do so.

"There is no one in the house but children, *mazboot* [right], Khalil?" the captain asked, all pumped up with adrenaline.

"No one but my family," my father assured him.

After they moved all of us into the living room, they pointed their guns at my father and ordered him to walk with his hands above his head. He ignored that and acted as if he were a free man in his own home. Most of the soldiers had fear sparking in their eyes, and I always wondered, what if one of them loses it and pulls the trigger by mistake?

"Upstairs, upstairs! Now, Khalil, we go upstairs," the captain demanded. They stormed into every room in the house with my father in front. The soldiers were worried about a trap waiting for them. We never knew if they really expected to find anybody, or whether it was just a drill. We were all frightened until my father came back downstairs safely to join us.

While he was gone, my mother kept asking the soldiers stationed at the door of the living room what they were doing with my father. They just yelled at her, "*Oskoot* [shut up]," and told her nothing.

For the most part, the soldiers avoided speaking to us. I think I now know why: they did not want to go head to head with my father. He prevailed in every discussion that arose, especially when they started talking about why he must leave the house.

"Would you leave your house?" my father would ask the soldier.

"No, I would not leave my house."

"So, I am like you. I will never leave my house," my father said, as if he had already known the soldier's answer to his question. "This is my home. This is my house. This is my land," my father would say each time, like a mantra. "It is my childhood. It is my memories. It is my family. The love of my land runs in my veins."

"OK, now you go back to the living room, Khalil," the soldier ordered my father when the inspection was completed. When he had joined us, they shut the door and locked us in until further notice.

One night, when the soldiers had nothing else to do, they did a search of the ground floor bedrooms, with my father in front as usual. They were always searching through our stuff, looking for anything they could use to make a problem. One night they found a notebook in which my little brother Mohammed Salah had drawn a missile hitting something that looked like one of the guard towers behind our house. They shook it in my father's face.

"Khalil, you said your family was for peace. Look at this!" They acted like they were personally offended. They were even angrier when they went through Mohammed Salah's computer and saw that he had watched a three-minute video online of some guys with covered faces firing rockets with loud patriotic Palestinian music in the background.

It was like they finally had something to put on my father, who spoke of nothing but peace. Truth be told, it was during those times that I saw what my father was all about. Every day they were treating us in all sorts of horrible and inhuman ways, and suddenly they were upset about a drawing.

Another time, one of them pulled open my drawers and dumped everything that was in them on the floor: my albums, my money, and all those letters and poems I had written about love. My father saw the letters and picked up a few.

That night when we were locked in the living room, he asked my mother, "Did you know that we have a son who is a great poet?"

My mother knew him well enough to sense when he was setting her up for a joke.

"If I have a son who is a poet, I would like to hear some of his poems," she replied. Everybody knew my father was talking about me. Yazan was too serious to write poems and the twins were too young. I just wanted to crawl under a *freash* and pretend I had a horribly contagious disease so that everybody would stay away from me.

"We are very lucky," my father said, "because I have one of the poems right here." He picked up one of my letters and started reading it out loud. My siblings all gave me sideways looks that were between pity and gloating. Then my father read another and another until he had read them all.

I had been busted. All of the love letters I had written were now in front of him like dead sardines. I got down on my knees in a desperate attempt to put things back together before the soldiers uttered another order. I was embarrassed. My siblings teased me. My father gave me a hard time about not paying attention to my schoolwork. I was drowning in tears inside.

It was a long time before I could laugh at the memory of that evening. It took an army, though, to find me out.

* * *

The crew of soldiers who locked us in changed every couple of nights. We never really got to know any of them. It was like we had to train each new group that came. Some knew how to behave, and some were just mean-spirited. The worst ones were from New York. They were loud and barbaric. As always, my father greeted the soldiers as though they were his guests.

"Where are you from?" he would ask politely.

"New York. Brooklyn. Y'ever heard of it?" they would reply, like they were trying to start something. "We left our homes to come fight," they taunted.

"Where are you from?" they asked arrogantly.

Sharply pointing his index finger to the floor with the most uplifting smile on his face, he responded, "I am from here."

Ever since then, I have had a bad feeling about New York.

There was one soldier named George who was an Israeli-Christian Arab. As soon as his captain saw that he treated my father politely, he was removed from the team. Most Arabs refused to serve in the army. A few Druze and Bedouins, however, were among the soldiers. They pretended they did not speak Arabic, but we knew they did. The Bedouins were valuable to the army because they could track footsteps in the sand.

Some nights, when the soldiers did not come for a while, we sat on the veranda and watched the bullets flying back and forth, like red dots speeding past before disappearing into darkness. Occasionally, the soldiers shot up a flare that was bright enough to light up a football stadium and lasted ten minutes.

When the living room door was left open, I carefully observed each soldier's speciality. Some had M-16s, or a massive sniper gun, or a bazooka. Sometimes they brought a German shepherd and teased it with a tennis ball. Whenever my mother saw the dog, she spoke to them like she was speaking to one of us.

"Make sure I don't have to clean up after that dog," she said with a tone of annoyance. Many times, though, she did.

The soldiers themselves were not clean. Sometimes we wondered if they were using our top two floors as an open toilet. Smells were carried down the stairs and through our shattered windows right into the heart of the house by the breeze coming from the sea. We were always debating whether they had done it on purpose. With time, we just pretended that we smelled nothing.

They threw candy wrappers, empty potato-chip bags and yogurt cups on the floor of the hall and the kitchen, on the stairs, outside on the ground. One time I saw a chip bag floating down from upstairs with a logo of a baby, an innocent smile on its face, wearing only a diaper. The image drew a wry smile to my face.

There was one red wrapper for chocolate that had a picture of a cow on it. I hated the untidiness of the discarded wrapper, but because I love good chocolate, I was very curious how it tasted.

We thought that they were just exaggerating their rudeness to antagonize my father and drive us from the house. Good manners to my father were as important as material wealth, if not more so. The impression they left was, "This is how these people live."

One time, some of the soldiers circled around my father. One began shaking him back and forth, asking loudly, "Why don't you leave this house, Khalil?"

My father answered sharply with his eyes zoomed in on the soldier and politely said, "Why don't you leave my house?"

"You should not be staying here," the soldier shot back.

My father replied softly, "I am a peace lover. I don't attack, nor hate, nor plot. Nor do I lose my right to exist."

I think that is why they never got to my father, because he always thought of them as children and always knew what to say to them. He dealt with them in a wise and dignified way. A man could either look down on the soldiers or be utterly terrified by them. My father respected everyone, but was terrified by nobody.

It took time to work this out in my head. Then, one day, I saw some of them drinking chocolate milk. This became a regular occurrence. Of course, they just threw the empty milk boxes on the floor when they had finished. Still, it cracked me up. They had these massive guns, wore helmets and flak jackets that made them look like comic-book superheroes, and drank chocolate milk. They really were just kids after all.

* * *

With so many truly bad things going on, the one thing I really hated more than almost anything else was when my mother sent me into the city to get the food coupons that came from donor agencies in other countries. It was so humiliating to stand in line with people who were not dressed well, with their teeth looking brown, their sandals barely holding together, and their shirts with holes in them. I would be cursing to myself the whole time I stood there, because to me taking the coupons meant that we were poor. If I needed to feed myself, all I had to do was take a walk through our fields and pick something, or go to the animal shed to snatch some eggs. My mother, however, nagged me to go get the coupons, then cash them in for the flour, milk, or cans of tuna she felt were rightfully ours.

"Go, go, we are suffering, too. Go get them," she insisted. "Why should others get and we don't?"

Each time I went, I prayed it would be the last. I did my very best not to let anyone see me, especially my friends from school. I felt God was punishing me. My mother thought that some justice was being served whenever we got the coupons. I thought it was a total injustice and I was paying the price for nagging her to send me into Deir all on my own.

* * *

One day I heard the words "buffer zone" for the first time. Apparently, the army had decided to create a protected area around the settlement. A few days later, the bulldozers roared into action in the afternoon and demolished the palm trees we had behind the house, including the rarest of them all, which produced yellow dates all year-round, so sweet they made my mouth water. The remaining orange trees were uprooted. The beehives got smashed. The bees that survived flew off in several swarms, while those that had been killed were crushed into the newly carved-up soil. Nothing had ever made my father more content than having a spoonful of honey from his hives and olive oil from his trees every morning for breakfast. Now even that had been taken from him.

Another day, the soldiers knocked down our red metal gate, my barrier from the outside world. Its loss left me feeling exposed—a feeling I had never felt before, not even with the soldiers all over our house. The bulldozer pushed the gate all the way up the driveway and left it in a crumpled mess, like a ball of useless paper, at the end of the courtyard where we had once sat with our guests.

The soldiers had demolished many of our neighbors' farms. My father was worried that they would destroy our greenhouses. He talked to my mother about our savings and how much they had or did not have. I had never heard my parents talking about money before. My parents could have no private talks about such matters, not with all of us locked up for so many nights in one room. These conversations made me feel very uncomfortable, but helped me to understand, in a way I never had before, that it was the farm and the land that had made our good life possible. Hearing them made me yearn even more for independence and freedom from the brutality of the Israeli occupation.

The soldiers would come and leave letters for the owners that their houses and greenhouses were to be destroyed at a certain hour, usually the next day, which left little time to save their contents. The letters would be signed by some major colonel whose name no one had ever heard before.

We never received any letter. The bulldozers took my father by surprise as they began demolishing our greenhouses in the middle of the night. We could hear the sound of the plastic sheets ripping and metal frames being crushed, all ten rows of them. My father had no one to complain to or argue with; the bulldozers simply could not be stopped. My father stood with us and put his arms around his waist and mumbled, "God, give me recompense."

"What recompense?" I asked loudly. "There is never going to be any recompense."

"If you succeed in your studies, you will give me recompense," he told me in a stiff voice when he heard my outrage.

And just like that, we lost the greenhouses. The next day we found a letter from the Israeli army outside on the ground in front of the house informing us they intended to demolish our greenhouses. My mother wanted my father to sue the soldiers, but he felt it was a waste of time. This would be the third time that he had sued the Israeli army. You could almost feel sorry for the poor judge having to decide three lawsuits filed by the same person. I thought my father should take mercy on the judge and just move on with his life.

* * *

Whenever I looked at the wreckage of the greenhouses, I thought about all the times I had pretended not to hear my father when he called me to help him on the farm. The work was not the problem. I just wanted to be the one deciding what I did.

Despite the madness going on around us, my father still demanded that I study hard and not use the situation as an excuse.

"When you want to, you can do amazing things, but only when you want to," he said when I did something that made him pleased with me.

School exams in Gaza are very competitive. You are tested only once in each subject at the end of the year, so you have to know every small detail of the nine subjects you are studying—Arabic, English, math, chemistry, physics, science, algebra, religion, biology—or you end up on my very competitive family's wall of shame.

Some students did not care, but most did. Everyone worked as hard as they could because they knew that the better your marks, the greater your chances of getting a scholarship to study abroad, which would set you up for life. The only way to live your own dreams was to study abroad.

One day I was studying in the same room as my second oldest brother, Yazan. It was during his final year of secondary school. He was being pushed hard by my father to get high marks so he would be accepted by a good university in Germany and become an engineer.

Suddenly, we saw the soldiers running behind the kitchen and setting fire to the baby green palms and the animal shed. Just like that! They were throwing buckets of petrol which caught fire and then exploded, creating a bigger whoosh of fire. My father went out and stood behind them asking over and over, "Why? Why? Why?"

They did not answer, but pushed my father out of their way as they retreated to their base. Yazan ran fearlessly to put out the fire with a bucket of water. As he did, a soldier in one of the watchtowers shot him in the foot with a rubber bullet.

He fell to the ground screaming and started crawling back to the house. It was shocking to see my tough big brother stumble and plead for help like a little child. My father ran to him, picked him up, and carried him back to the house. Then we carried him outside to the road to make it easier for the ambulance to pick him up.

During all this time, my mother was shouting, "Yazan, you should not have gone outside. Now you will have to be in the hospital and you won't be able to study." My brother just looked at her and did not reply. Meanwhile, my grandmother Zana was fussing over him. My father went with him to the hospital in Deir.

Somehow Yazan managed to live under occupation, deal with the soldiers daily, and still get good marks. He stayed in the hospital for two weeks and limped badly when he came out. That did not stop him from taking his exams and passing them.

The same was true of my older sister: she refused to be intimidated by the soldiers and was very creative about how she handled the situation. She wrote letters to them in English about peace, hoping they might become disillusioned with war and treat us more kindly or perhaps even leave us alone. She was very smart and received almost perfect scores on her school exams. She spoke English to the soldiers in the strong British accent that was taught in Gaza.

Once she made a sign with the word "jail" written on it, and hung it on the living room door so they would see it when they came in each day or night. She made slogans about peace and hung them around our living area. She even made the word "peace" out of some empty bullet casings we had collected.

Collecting bullet shells had become a hobby of mine. Rubber

bullets, real bullets, bombs, grenades, missiles, even tank bullets—I collected them all. It was like we had found gold whenever any of my younger siblings or I stumbled on unusual-looking objects that the soldiers had left behind. We would run to show our find to my father, who would shout at us, "Do not touch those things! They are dangerous! Don't bring them into my bedroom!"

Sometimes the missiles the soldiers fired at the house did not explode, so we had to wait for the Palestinian police—when they were finally allowed to come—to disable them.

* * *

Journalists were now starting to come to our house, and I liked showing the bullet shells to them. One day, the soldiers shot at the balcony attached to my parents' bedroom. My father had insisted I study there so that he could keep an eye on me and make sure I was concentrating. Thankfully, I was not on the balcony when the shots came, but my English book was. It got a bullet hole through the middle of it. I was ecstatic: I finally had evidence of why it was impossible for me to study in the house. The journalists certainly bought it, but not my father. Our situation had become international news, so journalists were frequently requesting permission from the army to interview my father. Media people were about the only ones who could get permission to visit.

My father always said the same thing to them: "The soldiers told me I should leave, but they only wanted the top two floors as an observation post. If I leave, I know I will never see my house again, so I am determined not to make the same mistake made by so many Palestinians over the years."

On one occasion, two audacious female Greek journalists were visiting. Sometimes they would come to the house, not to do interviews but just to hang out. That day the soldiers showed up early, before sundown. Without explanation, they confiscated the women's passports, and locked them in the living room with us. The reporters got to see for themselves what it was like to be imprisoned. The younger of the two held my little sister Zana in her lap and looked worried. Ten hours later, sometime after midnight, the soldiers ordered them to leave. With tears streaming from their eyes, they became very emotional, as did all of us, except my father.

Our imprisonment was hardest during summer holidays when we were off from school. I craved going upstairs and looking at the sea. My family lived so close to the Mediterranean that I could see it from the roof of our house. It was a magnificent masterpiece of white and blue with its waves, the shore, the birds flocking toward the horizon. I loved playing in it whenever we went to the beach during the summer, though we did not go often. Even so, the sea was always there. It was an important part of my life. I saw it whenever we were driving on the Al Rasheed Road that followed the shoreline.

In summer, the whole sky would be filled with colorful kites, even with the Intifada going on. I could see a few kites from ground level, but I wanted to have a larger view of the area to see how many were flying. Of course, the soldiers would not let me go up onto the roof.

Meanwhile, my father wanted my brother, Yazan, who had finished his secondary school exams, to leave for Germany quickly, while it was still possible for him to get an exit permit from Gaza. We never knew whether things would change and permits would be withheld.

About the same time, my mother told me about a relative of hers who had gone to a summer camp in America and ended up attending boarding school there. My head began filling with ideas about boarding schools, though my father was expecting me to finish secondary school at home and then join my siblings at university in Germany. I did not like that plan. Could this camp be the answer?

For the first time, an idea came crawling into the back of my mind: "Yes, this is my paradise. But I do not want to be here anymore."

7

Occupation

Everything in my life was now entirely out of my control. It had been that way since the soldiers had taken over our house. What I had thought we would have to endure for a week, or maybe a month, had now gone on for more than two years. I felt a real sense of urgency to get away. I was convinced that if I did not act quickly my future and my life would be doomed.

The soldiers had destroyed most of the trees with their massive bulldozers. I hated those monsters. They were so slow and ugly, not at all like a jeep or a tank or that fast little thing, the Humvee, that looked like a tank but was not one. Then the soldiers started shooting at our animals for no reason.

They hit our donkey one day with a bullet that slowly penetrated her body. She was shot in the leg, but the bullet came out of her neck a week later. It was only then that we realized what had happened to her. That poor white donkey. My father had kept her out of respect for his father. Over the next few days, we watched her die slowly. My grandmother was heartbroken. The donkey was one of the last things that connected her to my grandfather.

Next, the soldiers destroyed the animal shed. My grandmother decided to bring a box of chicks onto the veranda. She was bent over carrying the box, as if it were very heavy. I saw her struggling and went to help her carry it. It was as light as the feathers on the birds inside. I liked doing things for her.

The soldiers shot everywhere. They shot at the air. They shot at the house. They shot in all directions, sometimes aiming at the palm trees. Palm trees are like humans. They can live for a long time. The only way a palm tree can die is if its heart stops. The soldiers would snipe at the hearts of the palm trees. After just a few days, the trees would dry out and crumble.

One day I watched the soldiers shooting at sheep for fun. They would fire three bullets. The first one would land to the right of the creature, the second to its left, and the third would be a direct hit. With the greenhouses wrecked, the farms had become open fields. The wild grass had grown tall enough to cover the crushed greenhouse frames and was inviting to passing herds of sheep and goats, especially after a good rain. The Bedouin youngsters who looked after the herds were not from our neighborhood. We did not know them, and they did not know the soldiers' rules, so the soldiers would sometimes shoot at them when they "trespassed." It became almost normal to watch. I would sit in front of the house and look at the sunset, listening to the soldiers shooting randomly at things. They were probably as bored as I was, but I had to learn to ignore it.

* * *

With nearly all of our greenery wiped out, my grandmother Zana had to sit in the courtyard to get a bit of shade. That was the only place where the soldiers let us do things like play football.

"Yousef, where am I going to sit?" she would ask when I tried to move her from her preferred spot under the window, which just happened to be at the corner where I used my shoe to make a goalpost. I let her keep sitting there, and instructed my siblings, "Don't shoot at Grandmother." Sometimes, though, a ball would come speeding into her legs or land in her lap. My father would find out and get upset.

There was one olive tree near where she sat that had not been completely savaged by the bulldozers. Olive trees are tough. They can live for hundreds of years, as I learned when one had to be cut down for some reason and I counted the rings. My grandmother was nurturing new growth from it. We had to be careful not to hit one of the new shoots that were bravely growing on the old trunk.

I made up rules. If anybody hit the windows with the ball, the other team got five penalty kicks. If they broke any branches off the olive tree, they conceded ten penalty kicks. If anybody hit my grandmother, that was worth fifteen kicks. My siblings got into arguments with me about the number of penalties and just quit, leaving me angry that I was the only player on the pitch. They asked, "Who are you to decide?"

That was the problem, I thought. I had no power to decide anything in my life. My aspirations for independence were boxed in on every side. I wanted to engage with what was going on around me instead of waiting for my father to tell me what I could and could not do.

* * *

Complicating everything further, our house was now not only like a military base, it was becoming a broadcasting studio for international news organizations. Reporters were coming almost every day, sweeping past where the red gate had once stood. I thought they looked so cool in their jeans, hunting shirts and blazers that had lots of pockets but no shoulders. They represented all the major newspapers, magazines and radio and television networks around the world: BBC, ABC, MSNBC, Al Jazeera, Monte Carlo, German TV, TV5, French TV, and Israeli TV and newspapers. Whenever they met my father, I could easily see how impressed they were by him. He was not the kind of man they were used to meeting, but a man who had managed to elevate himself above the hatred, above all the violence, and who was preaching peace and tolerance.

He spoke to all who came regardless of the politics of the news organizations which employed them. He spoke in such a way that everyone was able to engage with what he had to say without becoming angry. The Palestinian reporters loved my father for his wisdom and his passion for a free and unoccupied Palestine. There were times when it seemed like some of them came just to talk with him, as much for personal guidance as to conduct interviews.

Media people are a lot like soldiers—presumptuous, intrusive, and demanding—but they carry cameras, notebooks, mobile phones, and smiles instead of guns and grenades. Another difference was that they always asked my father's permission to come into our house before they entered. At first, we found their visits exciting, but over time they became a nuisance. We had no control over who might arrive or when.

They always gave me some hope, though, that my Palestinian people were not alone. Perhaps the journalists would help their people understand that in Palestine there are people like my father who want to live in true peace. If enough people in the West heard about us, I thought, perhaps they would re-evaluate their support for Israel's army. Everyone would win if we got to live in peace, I thought.

I wanted to say this to the journalists, but I was too shy to speak in English and on camera. At school I would hear the other students talk about how the West supports Israel no matter what they do to us, but when I saw the journalists that was not as clear. "If the West is already against us," I wondered, "why do they send their journalists to interview my father?"

Each time he spoke to the camera in a way that filled viewers in other countries with respect for him, and maybe even awe. Many of them wrote to him. They seemed deeply touched and inspired by his determined belief in peaceful coexistence. He was a man passionately defending his right simply to live in peace in his own home.

Letters came from all over the world. My father received more than two hundred thousand of them from ordinary people in so many countries, admiring him for his stance.

My father told a Canadian journalist with the *Toronto Star* that the letters helped sustain us. "We felt so isolated," he said. "Then the letters started coming. And you find out that there are people all over the world who care enough to write. We are very grateful."

One visit from CNN remains memorable, because it created so much drama for all of us. While the camera operator was kneeling on the living room floor reloading her camera after an interview, she asked, "Khalil, may I go upstairs to take some shots? Do you think it would be OK?"

My father replied in his usual gentle tone, "As far as I am concerned, of course you can. But the soldiers have told me that no one is allowed upstairs."

She started up the stairs anyway, announcing in a confident tone, "I'm going up. I'm an American." She obviously thought that that would make everything all right.

Well, she was wrong. Several soldiers came rushing down the stairs with their weapons pointed at the camera operator and at my father and us. Moments like this were scary. You never knew when one of these men might fire. Other soldiers came in from outside and ran through the house like mad dogs. They were very angry and demanded that the camera operator hand over her tape. They jumped all over my father. One of the soldiers grabbed him and started shaking him, shouting, "Who allowed her in here? Who allowed her in here?" The cameras were still rolling.

Standing up tall, my father quietly responded, "I understood that she had a permit from you."

It always amazed me that he could talk calmly and stand confidently, as though nothing disturbing was occurring. I could never have done that. I would have cried for help or attacked the soldier for disrespecting me. I never understood how my father managed to stay so calm. It drove me insane. Although it was the soldiers who were creating our hell, it was my father with whom I was angry and confused. I saw it as his fault that we had to endure all the indignity. I begged him all the time, "Why can't I go and live at Sitie's house for a while?"

"Your house is where you sleep, Yousef," he would state in his usual solemn tone. There was never any room for debate.

The CNN crew was forced to leave. The soldier in charge told my father to never ever again let anyone go upstairs without his explicit permission. He was very clear: "From now on, this house is divided into three areas: Areas A, B, and C. A belongs to the Israeli army, B is yours, and in C you have to ask me before anyone is allowed to enter." Area B was the living room where they locked us in at night.

The soldier was short and my father guessed that he was barely in his early twenties. All of them had so many weapons it was as if they were drowning in their helmets and gear. With that, some went back to their base while others retreated upstairs. The soldiers were extremely upset, and so were we, but my father was not. It was as though he had no capacity for fear. I am sure this served to help me feel safe in a way. Even if there was shooting outside our door, all he had to do was hug me and I would feel protected.

A few hours after the CNN crew had left, however, my sense of security was totally shattered. Darkness had chased the daylight away, and we were in the living room. As usual, I was watching a movie while doing my homework. My mother never liked it when I did that. I could count on her to chide me, "You can't study and watch a film at the same time and expect to learn anything." Her words never seemed to have much effect on my behavior, though.

My father had gone to his bedroom for some quiet until the time he would be locked in the living room with the rest of us. He needed a place to read and write. He was working on an MA degree in English literature that he had commenced after the soldiers had moved into our house. It was his way of demonstrating to us, and especially me, that if one is committed to a noble purpose, it can be achieved even in the midst of chaos. His thesis was on Victorian attitudes toward women in three novels by Margaret Oliphant. He was examining the constraints facing women in an industrializing Western society from a century before.

He lay on his bed with a book in his hands, his head propped up on a pillow against the headboard, just below the window facing the tower behind his bedroom. It was a hot and humid night, 24 April 2001.

A cross fire started outside, and whenever that happened my younger siblings would rush to the windows to watch the bullets as they zipped back and forth, as I had done the first time it happened. It was exciting for them to watch the flames, though I always feared that a random bullet might hit one of them.

"Hey, hey, get away from the window!" I shouted at them, being the big brother. Usually they did, but slower than a turtle.

It did not take long for us to realize that the bullets were aimed at my father's bedroom. My mother ran to see if he was all right. Suddenly, we heard her shouting. We raced to his room and found him lying on his bed, with shattered window glass everywhere. The wardrobe was in pieces, and smoke obscured everything. Parts of the ceiling had collapsed on his head. He was all bloody and wrestling the fallen lumps of plaster with his hands. Shrapnel from some kind of grenade had lodged in the back of his head, which was bleeding profusely. For a moment, he looked like a crippled, weak man. When we tried to lift him, he refused our help. Instead, he attempted to stand up straight, with his hand tightly pressed against the back of his neck.

My mother called for an ambulance, but in the continuing cross fire no ambulance could safely come to our house. The closest they dared to reach was our neighbor's house about half a mile up Mekka Street. My brother Zaid, who had been begging to get his license but did not yet have one, showed he was a hero. With the bullets flying around them like wild locusts, Zaid guided my father into the Opel. Then he jumped behind the wheel and gunned the engine, determined to get our father to the hospital and in time to save his life.

I could not imagine life without my father, no matter how much we clashed. It was terrifying. Despite how much I had rebelled against his rules, despite my lack of attention to everything he tried to teach me, despite his "stupid" philosophy of peace, I was devastated to have him gone, even for one night, as was everyone else in the family. Each of us retreated to sit silently in his or her own spot in the living room. Now I found myself wondering for the first time about losing my father, and whether he was going to live or die. Either way, I knew that I had to be prepared.

If his head had been one inch higher on his pillow, my father would have been dead. By chance, he was lying just below the windowsill, which had protected him from the full impact of the bullets and whatever kind of bomb or grenade had injured him.

By the next morning, my father had already had someone call the school to have me excused from class so I could stay with him. I always felt important when I received permission to leave class, and that time he asked for me specifically. It seemed as though I was the only one capable of handling whatever emergency might come up, though it was probably only me who thought that.

When I got to the hospital, my father's head was wrapped in bandages, but he had an easy smile on his face. "They tried, but they did not get me," he said to me. I sat on the bed by his feet and immediately he reached to rub my face with his hand. In fact, he rubbed until my face was raw, smiling incessantly.

The same CNN crew who had stirred up the soldiers the day before came to the hospital for another interview. As the camera began recording, a reporter, Ben Wedeman, sat by my father and asked him, "Khalil, after what happened to you last night, do you still believe in peace?" My father's laughter and smile faded.

"What happened to me last night makes me believe even more strongly," he said seriously. "We must not let our anger continue to divide us. I will always believe in peace."

I stayed by my father's side until everyone had left and, at age thirteen, I asked him the first real question I had ever put to him. I feared that if I did not ask him then, I might never have another chance. I had always wanted to ask him about girls and sex and smoking, and things like that, but had been too embarrassed and terrified to do so. For those things, I talked to my science teacher, Mr. Ayman, who was close to me and treated me like a younger brother. I confided all my personal affairs to him.

Now, though, I wanted to ask my father about what it had been like when he was a teenager, but that seemed a strange thing to ask of such a dignified man. So instead I asked, "How? How can you wish to make peace with the people who have tried to kill you?"

We talked and argued and talked some more about this during the week he was in the hospital. Whenever I challenged him with my observations, I immediately ran into the solid and unrelenting wall of his convictions. I would listen and, for a moment, be inspired. Then I went home to the soldiers and became angry again. His words sounded so good, but he was the only one saying them. They were having no effect on the people who were making our lives so miserable. How could he find peace when he was on a roller coaster over which he had no control? A roller coaster on which he could not say, "Stop and let me get off?"

Two opposing realities confronted me constantly. I simply could not understand how to maintain a sense of hope when surrounded by the sounds of death. The "bam-bam-bam," "tak-tak-tak," and "boom-boom-boom" often overpowered my father's gentle reassurances.

It became harder and harder to know what to think about my father or his philosophy. Although, it was the soldiers who were creating this hell for us, I found myself getting frustrated with him.

When I am older, I vowed to myself, I will show them that what they are doing to us can be done to them. When I am older, I will teach them a lesson they will not forget.

I decided that I had to know more about them. I read up on the *Yahood*. I watched a film called *The Pianist*. It was about a musician who survived the Nazi occupation. It broke my heart to see how people could be attacked and stripped of their humanity just for being Jewish. So why were they now doing the same sorts of things to us? I was sad that the pianist never got to marry that blond German girl, and touched by how the man she did marry was the one helping the pianist escape evil.

When my father finally came home from the hospital, with his strong and confident self in our midst again, we felt safe. We knew, more than ever before, though, how important he was to us. If I had not been his son, I would have declared him a hero and shouted his name in the streets. I was his son, though, and I had to live with the consequences of his ideals.

8

Ramadan

Each day, each week, each month passed, all impacted by the soldiers in our house. We did our best to hold tightly to the routines that were important to us. For me, one of those was observing the holy month of Ramadan, which is my favorite time of the year, along with Christmas; even though I was not a Christian, I still enjoyed the Christmas trees that would be put up by the Christians in the Strip, and of course I did not mind having some days off from school.

Though Ramadan is a time of fasting, it is also a time for food. I began every day by planning what we would eat that night. One afternoon, my mother asked me as I was cutting onions, "Are you fasting just to eat, or to be a good person?"

"I need to eat to be a good person, and I need to eat good food," I replied as I chopped some spring onions with a large, broad knife like those that the chefs used on TV. I felt like a professional chef.

I was a foodie. After the soldiers took over our house, I began to watch cooking shows on TV. It was a total distraction from the chaos around me, and a good way to pass the time before the soldiers showed up. I began pleading with my mother to make the elaborate multi-course meals I saw being prepared in studio kitchens. Mostly, she just said, "No."

"None of that is real," she insisted one day when I was trying to separate egg whites and yolks. "They are just like all the romance in the soap operas or the wrestlers on World Championship Wrestling."

"Whoa," I interrupted. "Now you want to trample on my wrestling as well as my food shows?" She laughed. She may have been right about the wrestling, but when it came to the cooking shows, oh, I insisted that was real.

"Those TV chefs have dozens of people you don't see making sure everything comes out just perfectly," she said. Maybe they did, but they knew how to cook and garnish a meal in the right way.

Still, my mother understood that I appreciated good food. I loved it when she asked, "What would you like to eat today, Mr. Yousef?" Here was my chance to come up with ideas for menus, and then have a lot of arguments about whether the food needed more salt or to be spicier.

During Ramadan, we always ate Arabic-style, sitting on the floor. In the morning, I gathered the *freash* on which we had slept. I would lay them very carefully in a perfect square for us to sit on when we broke our fast after sunset. Doing this made me feel like there was some order in my life, something to hold on to. In the center, I spread a *sufra*, which is like a tablecloth except there is no table. Then, during the day, I kept checking back to make sure that all the plates and cups and silverware were in their right places.

It was my job to go shopping for the special things my mother did not consider to be important, such as the aubergines, yellow peppers, and turnip pickles. My mother would say that no one other than me ever ate the pickles. But when they were served, they were snatched very quickly from the plate, which made me smile.

A traditional dessert we enjoyed only during Ramadan is called *gatayef*, but getting the *gatayef* pancakes was a tall order. I had to go to the bakery first thing in the morning because they would be sold out by noon. Everybody wanted them. There was always a long line and I worried about them selling out before it was my turn to order. I needed two kilos, about fifty of these small pancakes. Sometimes I had to wait four hours. The bakery owner's son was my father's former student. I thought that should give me some advantage. A couple of times he took me ahead of the others, but mostly he just handed me a number, as he did to all the other buyers who arrived at the shop.

At home, we filled the pancakes with a mixture of dried coconuts, sugar, crushed walnuts, and pistachios, then folded them in half and pinched the edges tight. My mother would bake them in the oven for about fifteen to thirty seconds on each side to get that brownish color. After they had cooled for a few minutes, we drenched them in sweet syrup tasting of rosewater or orange blossom water. The *gatayef* smelled as good as they tasted.

During the week, the soldiers allowed only my mother to go to the kitchen at night to cook for us. She used the stove there, but did a lot of the preparation in the living room where I could help her. On Fridays, though, anyone could go to the kitchen where the cooking usually began after the Friday prayers at the mosque. My father relished having us go along with him to pray. He always enjoyed appearing in public surrounded by his children, whom he believed would soon all be highly accomplished in life. I liked playing the role of a bodyguard. I always walked a step behind him as he climbed the stairs to the mosque.

Before the soldiers occupied the house, preparations for the big meal on Friday, the Muslim holy day, actually began outside the house, where we slaughtered whatever meat had been chosen: rabbits, chicken, ducks, baby pigeon, or lamb. We cleaned the meat in the outside sink, then my mother did the cutting in the kitchen. My grandmother picked the green beans and prepared them for my mother to cook. My job was watching *Tom and Jerry* cartoons on the veranda, and then running to the kitchen during the breaks to check on their progress.

The cartoons would take second place, though, to watching my mother and grandmother bake bread in a large clay oven in the garden. My grandmother prepared the dough, then handed it to my mother, who oversaw the baking. My mother would send me away when I talked too much, which she claimed caused her to make mistakes and burn the bread. Once she had said "Get out!" there was no going back.

After the soldiers came, that oven was crushed by their bulldozers.

My mother and grandmother built another. When the soldiers crushed that one, my mother and grandmother built yet another. And another, many times. The bread that came from those ovens was so good that every bite made me feel like an angel had kissed my face, just because I had been a good boy. Have you ever been kissed by an angel? Well, I have, countless times.

After a while the soldiers said we could no longer go outside to cook. We were restricted to the kitchen. So, I shifted my focus to presentation, which became very important to me. I would strategically place platters of okra, hummus, fattoush, tabbouleh and other good things on the *sufra*, making sure everything was as I thought it should be. My parents found this amusing.

"Yousef," my father said to me, "Ramadan is not all about the food. Not everybody has as much to eat as you do. You don't even eat some of these things you are always asking your mother to serve."

"OK," I told him, "so we can share it with the people who do not have as much."

For me, simply preparing the meal was a form of artistic self-expression. Eating was not as important. With the right kind of presentation, having my favorite foods made me feel like a prince from the *Thousand and One Nights* fairy tales surrounded by his loved ones. It was just the best feeling. I felt totally at peace and utterly loved. Plus, I knew everyone enjoyed all my preparations and enthusiasm, even if they did not admit it out loud.

Sometimes, in spite of all my planning, the kind of formal meal I had envisioned all day would be suspended if the soldiers decided to storm down on us right before it was time to eat. They might still allow my mother to cook, but the atmosphere I was trying to create would evaporate.

My favorite meal was *maqlouba*. My mother makes the best *maqlouba*, although Sitie claimed hers is better. To get my mother to make it, I had to work on her. If she was in a bad mood, it did not come out well. I had to make sure all the ingredients were in the fridge and the kitchen was clean. I washed the sink, the plates, all the silverware. I brought out the cutting board that she only used for *maqlouba*. I laid out the knives very carefully, with the smallest one first and the largest one last.

Then, and only then, I went to her room and said, "Yama, I think this is a good day to make *maqlouba*." That was when all the motherly sadnesses began to come out.

"I'm tired. I worked hard. I did all the cleaning," she said with a heavy voice, but I kept begging her anyway.

"Stop bothering me," she said. "There are soldiers in the house."

"I know there are soldiers in the house, but we still have to eat," I told her. And if I said it the right way and kept on saying it long enough, she usually gave in.

Before the soldiers came, I could go across the yard to my grandparents' old house where I raised chickens I had brought from my uncle's hatchery. I slaughtered two or three of them and stirred them in a pot of boiling water over an outside fire to get the feathers off. I never liked doing this. I then gutted and cleaned them. Having the soldiers in the house meant we had to buy chicken from the shops instead of raising our own. They never tasted as good. Regardless of where the chickens had come from, my mother stewed them while I brought aubergines, tomatoes, potatoes, and rice from the storeroom and laid them out on the counter.

"OK, Yama, now all you have to do is to stand here and work your magic." She put her hand on her hip and looked at me while I spoke with a big smile. The hand came off her hip and she started arranging the vegetables. Even as a child, I knew how to motivate people.

She caramelized the onions, aubergines, and potatoes. Then in a deep, round pan she would place a layer of potatoes, followed by aubergine, onions, and lastly the chicken that she had pulled off the bones. She covered all this with rice and spices and laid an overturned plate on top of that layer. The plate was there for only a minute while she poured the broth from the stewed chicken over it, allowing it to run off the sides and fill the pan. If she poured the broth without the plate, it would make a hole through the layers.

Once the broth had covered everything, she removed the plate, then put the round pan on the top of the stove and let it cook for about forty minutes. She always timed it to be ready at *Iftar*, the time when we could break our fast for the day. When it came out of the oven, my mother would taste it, then ask if I wanted to taste it as well. Of course, instead of sampling one spoonful, I would eat four.

"*Khalas, khalas* [OK, OK]. Are you going to eat the whole thing?" She pulled the spoon out of my hand, since we both agreed it was ready. Then she flipped the round pan over onto a big tray, which is why it is called *maqlouba*, which in English means "upside down."

Before she served the meal, my mother roasted some peanuts in a pan on the stove to sprinkle on top of our salad, which she served with yogurt. I always took my time before going to eat with the rest of the family, so I could stay in the kitchen and enjoy any clumps of the rice that had stuck to the pot.

Every meal ended with fruit, usually watermelon or guava or sabra fruit when it was in season during the late summer. Some people call it cactus pear; it has a very sugary center and has to be peeled carefully because it is covered with long, spiky thorns. My father knew how to hold it so he never got pricked by what was left of the thorns, which had mostly been cut off.

Until the soldiers came and my access to the kitchen was limited, I usually helped my mother clean up after we ate. If I ever complained, she would quietly say, "OK, next time, when you say 'come and cook this,' we'll talk then." I would give her a big hug, and start carrying the dishes to the sink.

The truth was that I did not mind being in the kitchen, since it was where we kept the radio. I could listen to songs while I washed the dishes. In fact, I would stay in the kitchen as long as I could to hear songs by Amr Diab, Fadel Shaker, Ragheb Alama, Najwa Karam, Hussain Al Jassmi, and Elissa. My mother worried, though, that all those love songs were the cause of my love letters and poems. She often pushed me out and finished cleaning up by herself.

During Ramadan, we got to see special Egyptian TV series made for the long evenings when families sat around eating. They were all about love and marriage and were full of Egyptian heroes spying on Israel. When my sister wanted to watch one thing and I wanted to see something else, I would tell her, "I made all the food today. You were thanking me two hours ago." That rarely worked. If we got too loud, my parents would turn off the TV for the night. So, we learned to negotiate.

At the end of each day, my father followed his regular routine: shower, pray, eat, and then lie down to read in peace—"Transport myself to another world," as he put it—until the time came for night prayer. Most of the time, we all prayed together.

If for some reason I did not pray, my mother would tell me ominously, "On Judgment Day, Yousef, no one will care about you. I want you to be in a good place." I had been terrified by the idea of Judgment Day ever since I had first heard about it when I was seven years old. I hated the thought that our family would all be separated. She seemed to understand that. As soon as she said "Judgment Day," I got up and started my prayers.

Early in life I was aware of the concept of God, and I thought that meant my parents until they began asking me to pray to God.

Nights during Ramadan were always too short. It seemed like we had to get up again as soon as we had fallen asleep so we could have *Suhoor* before the fasting started again at dawn. It was a simple meal: water, some cheese, *ful* [fava beans], maybe eggs. Most of the time I skipped it. It was not interesting to prepare. Once we heard the *Azan*, everybody stopped eating, and I would start thinking about what we would cook that day for the evening *Iftar* meal. As the *Azan* came to an end, we all lined up behind my father, who led the *Fajir* prayer.

At the end of Ramadan, we celebrated Eid by eating whatever we wanted, whenever we wanted. We dressed in newly bought clothes and my father would go around in the morning to give everyone a gift of money. Then it was up to us how we wanted to spend that money; some bought fireworks like myself, others stocked up on more sweets. Also, we ate *feseekh*, a dried and salted fish from Egypt. Gaza is influenced by the Egyptian way of life. They ruled the Strip for decades, right up until 1967. We still used Egyptian books when I was in the lower years at school.

Before the soldiers took over our house, waves of people visited us during Eid. My father always had a bull slaughtered so we could feed our guests and distribute the rest to other families around Deir who were less well off. Each year he also spared another bull to honor the soul of his father. I drove around with him in his car as we dropped off fifty kilos of beef in one-kilo bags at the homes of all of the teachers who had taught him when he was a kid. Some were so old that they could not remember him. He never forgot them, though. He sat in the car out of sight, while I delivered the meat.

I enjoyed playing this role, like Zorro or Robin Hood helping people secretly. My father told me to give the meat to whoever came to the gate and then disappear. He sought no credit or thanks. Still, I always told people who was giving them the meat. My father was not happy about that, but what could I say? They always asked me who had sent me.

9

The Bullet

Time was slowly moving on, and my father's hair was turning ever more silver. After the soldiers demolished the greenhouses, he went to an Israeli court to ask for reimbursement, but he did not get any. I think he sued only because my mother would not leave him alone about it. He never seemed to care as much about material things as he did about living in accord with his beliefs.

"We have been peaceful all our lives and we shall always be," he told one of the journalists who came to interview him. "All this violence will vanish. Let us break the ice and forgive one another so our children can have a peaceful life." This was the message he stated repeatedly to journalists like Amira Hass of the Israeli newspaper *Haaretz*, Ben Wedeman of CNN, Chris McGreal of *The Guardian*, and many others.

As I heard him express such sentiments over and over, I sometimes wished I had been born in another country. Palestine is the Holy Land, a land that is meant to be flowering in peace, but instead all I saw around me was destruction. I tried to tell myself that it was an honor to come from the Holy Land. Before my family were Muslim, we were Christian. Before we were Christians, we were Jews. Palestinians and Jews have always been one family. The geneticists have proven it. So, why on earth are we fighting? If some must fight, can they not do it away from me? Away from my life?

Since school was the only place I was allowed to go, I began to think more about how I spent my time there. I started to develop a daily routine: wake early, be the first to use the bathroom, dress, go to the kitchen for breakfast, do my homework, wake my younger siblings, and get to school in time to play games before class: football, volleyball, basketball, whatever the other students were playing. Our school competed with other schools and I wanted to be on every team, even when the coaches told me I was not good enough.

With nothing else to do, I started getting serious about my studies. Omar and Hossam found that boring. The whole of Ramadan had gone by without any pranks. I was losing them as partners in crime. So, I came up with an idea for the last class of the day before Eid, when we would have three days off from school.

"Buy some firecrackers," I told them. Firecrackers were always for sale around Eid. I gave them some of the Eid money I had received from my parents and sent Omar to get some. It was very important that my father never find out that I was involved, or the happy days of Eid would quickly become hell for me.

It should have taken Omar only a couple of minutes, but he did not come back for a long time. I was worried he had chickened out, and was spending my firecracker money on video games. Just as the bell rang for the second class, he showed up. I have to admit, Omar was smooth. If he had arrived earlier and walked in late to the first class, he would have had to explain where he had been. Instead, he just slid into the class schedule as if he had been there all along.

He gave me the firecrackers and Hossam and I checked them out.

"These are bigger than the ones I thought you were going to buy," I told him.

Omar shrugged. "That's all I could find." I shook my head; what could I do?

"Next lesson," I instructed them like a general preparing a commando operation, "when the Arabic teacher is writing his notes on the board with his back to us, you light them. I'll give you a signal." Hossam and Ibrahim agreed to help.

The teacher had handwriting like calligraphy and wrote slowly. Just as he started writing a long passage from some poet on the blackboard, I nodded to Omar. Seconds later, Omar, Hossam, and Ibrahim fired everything, leaving me with the rest of the class to enjoy the teacher's outrage and fear. He dropped to the floor like a dead man. The noise those large firecrackers made was huge. He must have thought the army was bombing our school, as they had done to many other schools in Palestine. When he saw the whole class choking with laughter, however, he understood. He got up and beat everyone except for Omar and me.

"Sir, would you like me to hand in my essay now or later?" I asked him, so sincerely.

"Not now, Yousef, not now. Right now I want to know who thought it would be funny to bring firecrackers to my class," he stated with fury in his eyes.

After we returned from the Eid break, my father sent his assistant to our class and announced that whoever had set off the firecrackers must come forward. If he did, he would be rewarded for his courage. I was smart enough to know that if I came forward my father would not only punish me at school to make an example out of me but punish me even more at home in front of my siblings. If I broke down at any time and cried, my father would hit me again for being weak. To cry was humiliating. I could not allow that.

I told Omar to go and confess that he did it. I knew that my father would go easy on him and not treat Omar the same as he would have treated me. Omar got up and went to the headmaster's office. When he came back, he walked through the door with a smile on his skinny face and said to me, "Your father suspended me for a week. Your father is the best."

He was so happy he did not have to come to school for five days, and I was so happy I had got away with it. I never told anyone about it, as I knew they would tell my father eventually.

Of course, I did get punished by him for other things, like fighting. I would shout at him, "Yaba, he started it. I swear it's the truth."

He would coldly answer back, "It does not matter who started it. You allowed yourself to get into a fight." My father insisted that I must never act out of anger, fear, or revenge.

My father was also comforting to me. Whenever someone in the city was shot, he tried his best to keep me hopeful. When life got particularly difficult, he would say it was a sign that things were about to get better. No matter how hard it got, he was absolutely sure we would soon see a "light at the end of the tunnel." It was always the same thing: "We must maintain hope and faith, and love our fellow citizens of the world."

He argued that war could not go on forever. To help me feel better, he tried to teach me to imagine life as I hoped it to be. Because the soldiers would not let us go to the beach, he told me to listen to the sounds of the ocean and imagine I was there. I wanted to yell back at him, "Are you freaking kidding me?" His idealistic talk only frustrated me.

* * *

18 February 2004: a week after my fifteenth birthday. I was hurrying home from school. As usual I stayed after class to play football with my friends. I was wearing my cherished AS Roma jersey. It was the first real football shirt I had ever owned—all my many Real Madrid shirts were fakes—and, according to my mother, a waste of money. Only knock-offs were available here in Gaza, but this one was original, a gift from an Italian director who was making a film about my family called *Private*. It had number nineteen on the back and the name of Walter Samuel, a defender at that club. Every time I put it on I felt important, and was sure it helped my game. Right now, however, I mostly felt hungry. I hoped my mother had prepared something other than leftovers for lunch.

As I got closer to the house, I could see a United Nations Toyota jeep parked in front of what used to be the red gate. Two men and a very attractive-looking blond woman with a British accent were sitting in front of the house on chairs, speaking with my father. That spot—in clear view of the Israeli soldiers in the watchtower beside our house—was now the only acceptable place where my family could receive visitors. As I walked past I could hear the UN people asking my father about the confiscation of our land by the soldiers, a familiar subject of conversation. I paid them no mind and went straight to the kitchen, where I saw nothing but leftovers laid out on the table.

"Yama, you told me you were making fish today," I complained to my mother ungratefully. I fancied myself a gourmand and was looking forward to one of my mother's special fish dishes.

"You enjoyed what you ate last night. You'll enjoy the leftovers today," she said, too busy getting everybody's lunch to be concerned with my disappointment. In protest, I decided to get my bike and pedal into town and eat something there. Even though the school day was over and I was supposed to remain in our house, I figured that the UN people would keep the soldiers and my father distracted long enough for me to leave. I was fast on my bike, and I knew that I could make it to town and back before our visitors left.

Then I changed my plans. Not only was the blond woman very attractive, she spoke in a British accent that I found particularly charming; I wanted an excuse to listen to her talk. So, I decided to stay. I also figured that this was a chance to show my English-teacher father that I cared about learning the language he was constantly urging me to study, study, study.

Just as I was getting seated, a soldier in the tower closest to us announced over his loudspeaker that the time allowed for their visit was up. *Leave the home! Leave the home!* His command was loud and urgent. The UN staffers looked up in surprise. They had been with my father for fewer than fifteen minutes.

Nonetheless, they obeyed immediately. We all stood up and they began walking quickly to their jeep, which was parked about thirty feet away. My father and I walked with them. He and I stood in front of the jeep as they got in. When the engine started, I raised my hand to wave goodbye.

The jeep had just begun to reverse when, with no warning, I heard the sound of a single gunshot. Since there had been no shooting that day in the neighborhood, I instinctively assumed that the shot was being fired away from me in another direction. In the same instant, though, I felt something knock me to the ground, like I was crumbling. I tried to get up but my legs would not move. I grabbed my stomach and called out, "Yaba, I can't move. Yaba, I can't move. Yaba." It felt like it was taking hours for him to notice me.

I was in shock and could feel no pain, but neither could I feel my legs. I did not understand what had happened. I ran my hands over my chest, stomach and legs to check what was wrong with me. The weird tingling sensation that was sweeping over me was something I had never before experienced. Then I suddenly understood that I had been shot.

My father turned toward me and, for the first time in my life, I saw fear and worry in his eyes. It was as though he, too, was about to collapse. The shock in his face terrified me as he bent over me. "What is going on?" I wondered. Everything around me seemed to be moving in slow motion. I saw my little brother, Mohammed Salah, kicking a chair in anger. My mother was running toward us. My grandmother was trying to run, too, but she was too old to move as quickly as she wanted. I tried to take off my AS Roma shirt and see where I had been hit.

My father lifted me and carried me to the UN jeep. I remember the driver asking in fear, "What the hell happened?" and the woman answering in outrage, "The soldier just shOt him! He just shOt him!" Even in my state of shock, I loved her accent.

The jeep took off and I sank into my father's lap. I was sure I was about to die and kept repeating the Shahada over and over: "There is no god but Allah; Muhammad is the Messenger of Allah." Everything within me and around me started to change. I felt like the world was slowing down on me. I felt so betrayed, but I somehow did not care about that.

"I am going to die," I said as I looked at my father. I confessed to him, "It was me who set off the firecrackers in Arabic class."

I can still hear him urging me to keep my eyes open. "Yousef, do not close your eyes. Keep looking up at the sky. Do not take your eyes off the sky."

"I'm sorry about my marks. I'm sorry about my English test. I'm sorry I didn't make you proud."

I wanted to do anything I could to make it easier for him. Sometimes we wait too long to say what needs to be said. I thought I was running out of time, and took some small comfort in the fact that I could still hear.

By the time we reached the hospital in Deir el-Balah, the pain hit me. Badly. I figured if I could feel that much pain I was still alive. So many strangers popped out of nowhere wanting to help carry me to the doctors.

"Khalil's son has been shot," they shouted. "Allahu Akbar, Allahu Akbar," they kept on shouting. My father issued an immediate appeal that no Palestinians retaliate in any way. For a second I relished all that attention in my usual way, but the pain was all over me. Pain was more cruel than the soldiers and as unrelenting as my father when he was demanding top scores. Pain had no mercy on me.

In my mind I kept seeing the footage of young Muhammad al-Durrah being shot and killed while the world watched on TV. I thought of his funeral, where thousands had come and chanted slogans. "No, no, no," I said to myself. "I don't want to die. I don't want a funeral. I want all those people cheering for me in a football stadium."

I hopelessly attempted to stop the doctors from ripping my AS Roma shirt in half, but they did that anyway. Then, as I was muttering angrily at them, everything went dark.

When my eyes finally opened, I could see people cheering. I could see my parents along with some relatives, who were all happy that I had just woken up. It was weird for me, though, and it was like I had never met them before. I was in pain and naked and miserable and all I wanted was to be put to sleep.

I do not know how long they had been waiting for me to wake up. It must have been a day or two.

"Why am I sleeping on my stomach? Why can't I move?"

"He shot you in the back, Yaba," my father told me softly, his head bent down.

The soldier had shot me in the middle of my back. He had fired from the tower nearest to me, barely sixty feet away. His bullet had left a hole so deep that the doctors could see through to my spine. I was paralyzed from the waist down and unable to lift my legs when asked to do so.

I started crying and crying. I thought that the world had given up on me so quickly, though I had never given up on it. "I am paralyzed and crippled," I kept saying to myself. I had such dreams and now nothing seemed possible. I kept a tight hold of the doctor's hand; I was in horrific pain.

"Make me sleep, please make me sleep again," I pleaded with his hand locked into mine.

"OK, let go, Yousef, I will make you sleep," he promised, but when I let go of his hand he walked away from my bed, leaving me burning in pain.

Day after day, my mother would hold my hand and say, "Be strong, Yousef, be patient." I hated those words. I was a total mess and no one could do a thing about it, yet I was supposed to be patient. My family could not even do anything about the fact that a soldier had shot me for no reason. I was used to the sound of raging cross fire with thousands of bullets flying everywhere, but this time there had been only one shot. One shot had caused me to fall paralyzed to the ground, and it had not been my fault. I had done nothing. I was intensely angry about everything.

To tell the truth, the person that I blamed most was my father. I was furious with him. I was the one who had wanted nothing to do with politics and peace. I was the one who had wanted us to leave. Yet I was the one who was now paralyzed. I would helplessly shout at him, "I told you to let me go live at Sitie's!"

My father never responded, as though he knew he had paid a price, a very heavy price, for not having been willing to leave the house.

As it turned out, the news of my injury received a lot of attention in the international media. Because the soldier had shot me with no provocation, because the UN had been present at the time, and because my father was well known for his beliefs, the story was reported everywhere.

My father told the journalists, "The soldier aimed to kill Yousef. There is a bullet in his spine. What else could he have been trying to do?" He said it was part of the ongoing Israeli effort to drive my family from our home.

"They know well that I am a civilian and peace-loving," he said. "I have never been any danger to them. On the contrary, I have always called for tolerance. I have experienced suffering at their hands. Every night they imprison me in my house in one room. Three years ago, in April 2001, they threw a grenade into my house and I had to be taken to the hospital. Now they have shot two of my children. In spite of all this, I believe it is time for tolerance. There is no time for anger. There is no time for revenge."

As I lay in the hospital, I could hear everyone talking about what had happened to me, but was in too much pain to understand. I became upset with the doctors and everyone and everything. I was hurt, I was scared, I was confused. No one was telling me the whole story about my situation. They would simply walk in, rub my forehead, smile, and say nothing. The truth was, there was little more that they could do for me, certainly not in any hospital in Gaza.

Everyone seemed to be trying to decide what to do with me. Someone would suggest that I should go to a hospital in Germany. Someone else would argue that a hospital in Jordan or Egypt would be better. I was wondering whether I would ever walk again, and others were asking whether my chances warranted the effort required to make that happen.

My head whirled with these unseen voices beside my bed. Usually I was too drugged to move my eyelashes and see who was talking. Sometimes, though, when they thought I was asleep, I could still hear them. I knew that, besides the crowd near my bed, another group at the door or out in the hallway was discussing what should happen to me. It seemed like everybody had turned into a medical expert. I tried hard to stay tuned to what they were saying—though it was hard to listen to their uncertainty and to have no say in what happened to me.

I could hear my father's voice the most clearly of all. He kept telling me not to worry, reassuring me over and over that I was going to be all right. I heard him telling reporters, "The bullet stopped in his spine. Every day I ask if he will walk again. The doctors don't know, but they say there are encouraging signals because he moves his toes."

Sometimes I would imagine myself as the cat in the *Tom and Jerry* cartoons. Tom was always able to get over any horrible thing that befell him. He might be crushed between two boulders, but then magically he would be restored to his former self. It made me giggle in my mind—I could never giggle out loud because of the pain—and for a brief moment I imagined being restored to perfect health. Soon, though, my body brought me back to reality. The stabbing pain was always there.

Then one day it seemed as though everyone I loved and some I did not love were there beside my bed, forming a sort of line to kiss my forehead one by one and say "Goodbye."

I wondered: *What do they mean by "goodbye?"* It sounded so final. Before I could figure anything out, my father and I were in the back of an ambulance. A short time later we arrived at the Erez checkpoint into Israel.

Part Three

ANGELS

"Repel evil with good, and your enemy will
become like an intimate friend."
—The Holy Quran, Surat Al-Fussilat (41:34)

10

Tel Aviv

As the ambulance waited in the long queue of vehicles trying to go through the Israeli checkpoint, I remembered the trip I had made to Jerusalem at a more peaceful time with my mother and Grandmother Zana. We went to Al-Aqsa Mosque to pray. I had always wanted to travel to foreign places, but I did not consider Jerusalem to be in another country. It is ours, too.

Even though it was a time of relative peace, we had been required to walk through a long tunnel before reaching some security cameras. Through a loudspeaker, the security agent told us what to do and what to show. We never saw anyone in person unless for some reason they wanted us to see them.

In the ambulance, I could not see any of that this time. I just lay there until it was our turn to pass through. I thought about the last time I had been to Erez four years before, when the Israelis had allowed my oldest brother to fly from Ben Gurion Airport to study in Germany. Palestinians could fly from Ben Gurion only with special permission. He was the only one of us allowed beyond the checkpoint.

The soldiers came and pulled my stretcher out of the ambulance to search it, the same way you would yank open a drawer if you were looking for a pair of socks or your expired passport or something unimportant.

"Please be careful," my father pleaded with them. "My son gets in pain if we move him." They ignored him and asked him where we were going. I presumed they were taking me to Ben Gurion Airport just north of the Strip and airlifting me to Germany.

107

The doctors in Gaza, though, had been afraid to have me moved as far as Germany until specialists could thoroughly examine the bullet in my back with all the best modern medical tools. They did not yet know whether the bullet had severed my spine, and were afraid that it might if I was jostled.

My father tried to make a joke. "Look at you, Yousef, you are already going places." I said nothing.

During the three days I had been in the hospital in Gaza, my relationship with my father had changed in some very important ways. He was treating me differently, more casually, as though we were friends. There had been absolutely no mention of homework or math skills that needed to improve. It was as though, because I was now paralyzed, I had been liberated from his high expectations, at least temporarily.

After a number of soldiers searched my stretcher several times, one of them slid it back into the ambulance with a bang. Though I was heavily sedated, the pain exploded all through me as if I had just been shot again. The ambulance moved out of the checkpoint and toward Tel Aviv. My father's German friend, Rudolf Walther, had helped arrange for me to be moved to the Tel Hashomer Hospital in Tel Aviv.

I felt so much confusion. It was as though, in some strange way, my wish to "get away" was being answered, but at what price?

When we arrived at Tel Hashomer, they took me first to the emergency room where they checked me over. Then I was rolled to a large room with two big windows. Although I could not move to look out, the room was spacious and welcoming. I felt very shy, though, and exposed; the staff kept taking off my clothes to examine me. Given the location of my wound, the blue hospital gown could not be tied. For weeks I had to lie naked under my blue sheets.

I hated that almost as much as being paralyzed. No one had ever seen me naked before except for my mother, and that stopped when I was about ten and I had fought with her about bathing me.

"I'll shower on my own," I would scream to her.

"You don't shower long enough, Yousef, and when you do you don't do a good job," she would answer back.

When she bathed me, she used a piece of fiber that she had picked from a fiber tree in the garden. Fiber trees grow almost anywhere. They spread out, covering everything in green like a tree from heaven. Anyway, my mother would rub me so hard that it would hurt.

During the week, every time I provoked her anger, she would just say, "*Tayib, tayib* [Very well, very well]. I will show you the next time you are in the shower." When the weekend came and it was shower time and I knew I had no choice but to turn myself in, I tried to reason with her, saying, "OK, just be slow. OK? Don't hurt me?"

"I don't know what you are talking about," she would claim innocently.

The nurses made me think of my mother when they were washing me, but they did not use a piece of fiber tree and they never hurt me. All the nurses looked so young and so beautiful, even the ones who told me they were mothers. I quickly found myself not caring whether they were mothers or not. They walked around the room with either blue or dark pink uniforms. I wished I could see them in the clothes that they wore normally.

I do not remember how many operations I had over the next four weeks. There were a lot. I kept seeing faces that were covered with medical masks. People held up their fingers and asked me how many were raised, or they asked me if I could lift my legs. My bed kept getting moved from one room to another. The pain was always there and I was taking all the drugs the hospital would give me.

I wanted to be asleep all the time. Sleep was pills, and wakefulness was pain. Every time I woke up, I was so scared and upset because I knew that it would just be moments before the pain hit me again.

I pleaded with anyone who walked into the room, "I need a nurse." Then when one of them came, I would grab her hand and beg, "Please, please make me sleep." Their hands were warm and soft and sometimes arousing, which made me realize in some hazy moment that *that* part of me was not paralyzed. Somewhere in the confusion fogging my mind I felt a small sigh of relief.

They attached two blue machines with tubes that pinched into my arms and a button I could click when I needed more morphine. I was soon clicking it every time I was even slightly awake. To keep me from moving, the nurses had tied belts around me and raised the sides of my bed. That made me feel like a creature from a movie that had to be restrained.

When the morphine hit, I floated in a world where I was a child who thought my father and mother were the creators, the founders of life, and the masters of my universe. Where my brain was like a crisp sheet of paper that had never been crumpled. No beliefs, no concerns, no contradictions, no questions. As I floated, I tried to hold on to that child. Sometimes I saw him wandering around the room of my thoughts and passionately attempting to stay visible. He is shaking and frightened by all that has happened. Yet, he seems to have accepted reality and wants only to protect himself. "Do not forget me," he gently demands. My creators had never planned on having me, but a sister who was one year older than me died. They decided to try to have another girl. I arrived instead. Perhaps that explains why as a child I was all too eager to live, too happy to care or to shy away from my mother's camera. I had brown hair that turned pure black as I grew older. I always asked to be dressed well, and enjoyed posing. I was eager to embrace life and life was eager to surround me with its silky, beautiful arms. My father was my protector: handsome, slim, and always smelling like he had showered in a pool of flowery scents. He dressed in dark-colored clothes: black, brown, green, and anything that was sharp and bold. I get close to him during his weekend afternoon naps to rub my face against his shiny beard before he clean-shaves and goes to teach English at the Sokyinah Girls High School.

The wound was the size of the base of a coffee mug. Its bandages had to be changed every morning and every night. That was awful. It did not like to be touched. I tried to tell myself that I had been infected by some kind of magical germ like the Goa'uld of *Stargate SG-1*, even though that comes out from the stomach.

My father took time off from his school and stayed with me in the hospital. For the first time in my life, it was he, rather than my mother, who was taking care of me. It was he, as well as the nurses, who would straighten my sheets, give me showers, and feed me. I was unhappy to have him see me so helpless. And, worse, I hated being naked in front of him.

I made my peace with the nurses seeing me naked all the time. I thanked God for the nurses. Something about them helped to cheer me and made me want to be a good patient for them. The nurses were always smiling. Even when I was naked and they were helping me with my bodily functions, they would be smiling. Soon I found myself smiling back.

I still felt very shy, especially when they would come and put a tube in my penis so I could urinate into a plastic bag. That was the most annoying of all my daily routines. It was painful when the nurse put it in and even more so when she pulled it out. She would look at me and say, "One, two," then pull it out quickly before she got to three.

"You need to get used to it," Seema, one of the nurses, would say with a smile. "Just once every morning, and when I need to change the pee-pee bag."

Seema was a Jew from Iraq. I decided that she liked me the most of all the patients. I loved her. She spoke good Arabic, but mostly we talked in English. She was always trying to cheer me up or get me motivated. She told me about her children and how she was working hard so she could send them to university soon. She often spoke with my father. She was a reader and they had lots to talk about.

I did not get to know much about the lives of the other nurses. They were the ones asking the questions, not me. I did learn a lot about their capacity to care. They quickly became the one good part of my very bad situation. They made me smile no matter how I was feeling or what I was complaining about.

Whenever I remembered that my nurses were Jewish, I felt confused. In many ways, they were just like the Palestinian women I knew at home—kind and very caring. Until then, I had only thought of Jews as soldiers who pointed their guns at my family and me. It was a Jewish soldier who had shot me, but the nurses were also Jewish. "This whole world needs to stop messing with my head," I thought. I had so many different ideas all at once bouncing off the walls of my brain. It was extremely overwhelming.

An old man named Alexander shared the room with me in the surgery ward. He could not speak a word of English, and when his family came to visit they only spoke Russian to him. They were friendly to me and my father. Alexander had a TV, and they let me watch championship football on it. His grandson Yacov talked to me. We talked about sports and of course some politics. It was all very strange to find myself having a normal conversation with a Jewish boy of my age. I gave him my e-mail, and he told me he would keep me informed about his grandfather.

Before I had come to the hospital, the only Jews with whom I had ever interacted were the soldiers and the settlers. To my knowledge, I had never even met an Israeli civilian before, although my father told me that Israeli traders used to come to our house to buy tomatoes from us.

"During the peaceful days," he said, as he shut his eyes and opened them again.

All I knew about Israelis was that they had guns and had the power to tell me and my family when to use the bathroom and when to go to school, and that one of them had almost ended my life a few weeks earlier. Apparently, just because he could.

When I talked to my father about the medical staff, he said to me, "They are going to save your life, Insh'Allah," and because everyone was so kind to me, I began to believe him.

They brought in a couch that opened up to be a bed where my father could sleep next to me at night. I had never seen one of those before. It was dark blue. I like dark colors. My father looked good sitting on it. He sat by my side day after day. After he had done all he could for me, he put on his giant pair of reading glasses and read. When I went to sleep, he would be reading; when I woke up, he would be reading.

I never told him, but as I watched him read, I found myself observing him closely and loving him and loving him. I thought about the times when, as a small child, I had gone to his closet to breathe in his clothes that smelled so good, and sometimes I even tried to put them on, though they were all too large for me.

When the nurses and doctors were not around, he and I would reflect on all that had happened and all that might happen.

Sometimes, he gave me one of his speeches about thinking positively. Sometimes, I did not want to hear all that again. I had been told that if he was tiring me I was to tell him to leave. I loved knowing I had that power. I never used it. I wanted to keep hearing his voice, and he would talk until I eventually managed to fall asleep.

The German ambassador paid me an official visit along with his two assistants. The ambassador sat down and shook my hand, but spoke mainly to my father who had worked with the Germans on several projects in Gaza besides the Rudolf Walther School. Secretly, I wished he had visited me in Gaza instead. Then, everybody in Deir would have been talking about it.

A day or so after the Germans visited, a few military men in dark green uniforms walked into my room. They had been sent by the Israeli army after they heard the German ambassador had been to see me. Right away I knew why they had come: they wanted to make a show to the German government that they were good guys, and were "sorry" that I had been shot.

They behaved very kindly, but I was not comfortable in the atmosphere they created in the room. From what I had experienced, I believed that the soldiers could never be kind, even when they had to pretend to be. They were so different from the nurses whose kindness was real.

"Why did he shoot me?" I asked them over and over. They did

not answer. "What did he tell you?" No answer. It was like they just had to be there.

We had heard through the media that the soldier who had shot me was a captain. The Israeli army issued a statement that the soldier claimed to have shot "in the direction of the wheel of a vehicle that looked suspicious to him." The only vehicle in the direction the soldier fired was the clearly marked UN car. They said they were investigating the incident, but they never came to talk to my father or me or the people from the UN who had been visiting our home.

The military men tried to apologize to me, but I felt their apologies were fake, and I would not accept them. For me, this was way too personal.

My father did accept the apology, because he did not want to stand in the way of any kind of goodwill. For me, though, it went beyond my father, his land, his values, and his entire world.

All I could think about was the soldier who shot me. Who was he? Where was he? Did he ever think of me? I cursed him every time I felt my wound, and the outrage rose up in me all over again. I spent hours imagining that some day I would meet him again so I could ask him: "Why?"

11

The Sheekum

After a month of surgery, my wound stabilized sufficiently and my pain eased enough so that my father could leave. He said he needed to get back to his school.

"Why do you have to go to school if the students are not coming?" I asked. I was keeping up with the news as best as I could by watching CNN on the hospital television in the hallway by the reception desk. I knew the Intifada was still happening.

"Yousef, we must never let other people get in the way of our education. If the students do not come," he said as he stretched out his hands, "that is their choice. But the school will not be closed. That is my choice." He cared deeply about his school and was very committed to making it a model school in Deir el-Balah and perhaps even the whole Strip.

For the next three months, my mother took his place. She would sit by my side and do things for me the nurses had no time to do, like helping me to eat or shifting the curtains if the sun was in my eyes. Sometimes when there was nothing for her to do, she would wander around and mingle with other Palestinian women at the hospital. The families of these women had stayed in 1948 and were now living in Israel. Their presence gave me a sense of comfort and hope. My mother always came back with stories. I listened to her as she told them, but I had a hard time keeping track of everybody.

Sometimes my mother helped me change my bandage, but I preferred the nurses. It was like when she took me to buy new clothes or gave me a shower. She had a suggestion and I had a different suggestion, so we just ended up going back and forth. Then I got in a bad mood and blamed her for it. For some reason, I was more open to being tough on my mother than I would ever dare to be with my father.

My mother and Seema got along well. They spoke Arabic together, with Seema speaking in her sweet Baghdadi accent that I loved to hear. They talked about their children, and all kinds of woman things, like what creams Seema thought my mother might like or what shampoos they used. I watched them closely. "My mother is a Palestinian," I told myself. "Seema is an Israeli. They are just talking woman-to-woman like they have known each other for years. Why can't everybody be like them?" All the nurses liked my mother. At night, when she slept on the same couch my father had used, they made sure she had everything she needed.

* * *

I still could not walk, and could only barely roll over in my bed. With the wound slowly healing, though, I was regaining some of my strength. Well, that was what they told me. I did not feel it. The time had come to move me to a rehabilitation center. It was in another building called Sheekum Yeladem, the Children's Department.

They got me into a wheelchair. Wearing my hospital uniform and my black flip-flops, and with my pee-pee bag in my lap, I was gone. On the way, I saw more of the hospital for the first time. It was not just a hospital. It was a whole complex of tall modern buildings connected by a glassed-in bridge.

As I rolled into the Sheekum lift, I did not know what to expect. However, when I arrived in Room 5, a colorful room with a computer next to my bed, I relaxed. I was really excited about the computer. Every child had one. Now I could play FIFA video games again. My mother made sure that I did not watch anything I should not.

The room had two beds, so I knew I would have a roommate eventually. There was even a sofa bed where my mother could sleep comfortably, though she never seemed to sleep much.

They brought in two wheelchairs; I had a choice between red and blue. I picked red, because it looked bigger and I thought it would be more comfortable even though I would have preferred the blue one for its color.

Dr. Amichai Brezner, the head of the Children's Department, was especially kind. He was a bald-headed man wearing a pair of round glasses. When I arrived, he greeted me with a smile.

"Welcome, Yousef. I hope you will feel at home here."

Home? His words made me think I was going to be there for a long time.

"Am I going to walk again, Dr. Brezner?" I asked.

"That's the plan," he answered noncommittally. "It is not going to be easy, but we will do everything we can to get you back on your feet." He made me feel like RoboCop. In the movie playing in my imagination, I was the good guy who was supposed to be dead but instead comes back to life, not just strong but really strong. I tried to feel optimistic.

Over the next couple of days, I started meeting some of the other children in the Sheekum. I was shy at first. They seemed a lot more "modern" than I was, and I knew they thought me "different." Some of them had their ears and noses pierced. My mother did not say anything about that, but I knew what she was thinking by the way she held her head when she saw one of them.

When families came to visit, everyone was free to go from room to room. Some brought goodies to share. Finally, I had a chance to taste some of that chocolate wrapped in the picture of a cow. Oh, I was amazed by that chocolate cow. It tasted so good. Every morsel was heavenly. Chocolate is definitely something from heaven, so every bite felt like it was working a cure on me.

There were both boys and girls on my floor, and one Israeli girl in particular whom I liked. I was told that she had been in a car accident and had become paralyzed. She had almost lost her larynx, too. When she coughed, she woke up the entire floor, though everyone was more than happy to go check on her. She was so beautiful.

I knew I did not have a chance with her. I was just a Palestinian teenager in a wheelchair, but I could fantasize. She looked like a queen sitting in a chair of glory. Michal was her name, but she liked everyone to call her Micha, which made it sound as if she were a boy. She was too beautiful to be a boy. Her short black hair was cut so splendidly it made her even more attractive.

"Do you want to see my tattoo?" she offered once, out of the blue. "Here, look here."

I felt a little shy when she partly pulled her shirt off to show me her back, where she had her tattoo. She seemed fascinated that I had been shot in the back. She always wanted to look at it and touch it. That almost made me feel OK that I had been shot.

Micha was hanging out with another boy on the floor. He had his eyebrows pierced.

"It's cool, man," he said when I asked him why he had done that.

He wanted to be called Miko. He was an Israeli, and the handsome kind of guy who won all the girls, but he was also fun. He liked to race around in his wheelchair. He would wheel into my room and shout, "Yousef, Yousef, let's go."

Mohammad was another kid from Gaza. He had been wounded while he was walking near a car that was bombed by an Israeli helicopter. He was learning how to walk again and would have to live with metal sticks in both of his legs for the rest of his life.

"For the rest of my life, for the rest of my life," he would say in a way that almost sounded like enthusiasm, but a strange type of enthusiasm.

Given the situation in Gaza, we were both happy in a way to be at the Sheekum. We could go anywhere in the hospital. There were no "military" rules, and we could play games like UNO and have fun everywhere. Some days it was so much fun that it seemed like we were almost glad we had been injured. I had reached the point where I could roll myself in my wheelchair without help.

Whenever I saw Mohammad, he was always laughing about something. I would stop being upset about my situation and start laughing with him. He would drive the nurses crazy during physical therapy. He always tried to do an exercise differently from how the trainer wanted it done, just to show off.

It was our job to get up in the morning and go do physical therapy. Miko usually came in late, and Mohammad, too. They were roommates, so they were always up all night for whatever reason.

I made sure that I woke up early and went there on time. First, because I did not want to deal with the nurses taking me to the bathroom, I would try to do that all on my own now. Second, because I really enjoyed physical therapy. It was fun. I guess I was still a child; all we were doing was playing games as if we were five-year-olds. They had us throw a lot of balls around to help us regain our strength and coordination. Miko, Mohammad, and I could barely shoot a small air ball so it was humorous watching each other fail over and over again.

Miko spoke only Hebrew and Mohammad spoke only Arabic. They each knew a couple of expressions in each other's language, like "How are you?", or in English, like "It's cool, man, it's real cool." Neither knew English well, so we usually did not bother talking about anything as much as we did things together. We did not need to talk so much as just run down the clock and get better. Miko told me in broken English that he had crashed into another car when he was speeding.

"Crazy man, craaaaaazy," he would say every time he spoke about what had happened to him.

We often rolled around a small shopping area in the hospital complex, which was near our building. Crazy Miko called it "the mall," and always won our escalator competitions when we went there. Miko had bought his own sporty wheelchair, but Mohammad and I just stuck to the big hospital models. Naturally, I really wanted to have a fancy wheelchair, too, but my gut prevented me from even expressing the idea. I did not want to accept that this was something I was going to need for very long.

"What difference would it make?" I asked myself. "I would win the escalator competition that Miko invented?"

When we raced across the bridge, Miko always won because he had the better wheelchair. Mohammad would demand, "Let's race again." I kept on rolling away, though, until I was separated from them and they had gone somewhere else in the mall. As much as I loved all the laughter, I found myself needing to be alone sometimes to think and to process the world I was beginning to discover.

Usually I would head to the shawarma joint to eat. With a mouthwatering shawarma sandwich, I could work things out more peacefully. The man who ran the shop was named Nicky, or at least that is what everybody called him. The meat was kosher, which was the same as halal, so I had no problems eating it.

Nicky was a middle-aged giant of a man who enjoyed having me show up to buy his food. I was very inventive about the kind of salads or sauces that I wanted him to put on them.

"Just the beef shawarma with the tahini sauce and onions?" he asked, "or one with everything? Or perhaps a shawarma with French fries and Israeli salad?"

"Arabic salad," I would interject. "Let's be more inventive than that, Nicky." I tried to come up with orders for things I thought he did not have, just to challenge him. But he always had everything.

"Hot pickled yellow peppers on the side, right?" He was impressed by how much I liked things really spicy. "And some fritters."

Sometimes I took my food and rolled away quickly, but he would shout my name with his Eastern European accent. "How about some pickles, Yousef, ah?" He said it the way a good cook would say it. I turned around smiling and quickly took whatever he was offering from his hand, eager to go and eat.

Once I rolled with my lunch into my room just as a couple were moving in their son. They looked so blond and beautiful, but I saw a weird look on their faces. I went to my side of the room and started unwrapping my food, and offered to share it with them. They waved with their hands, "No."

My father visited later that day. Whenever my father came into my room, he flooded it with his eternal enthusiasm. As soon as he saw the couple and their son, he greeted them warmly in English, then asked, "Anything we can do to help?"

"Oh, it's all right. We seem to be getting the hang of it here," the boy's father said with a strong American accent. I later overheard him tell my father that he was in fact from the US. My father would always engage in conversations with him about the Palestinian–Israeli situation. No matter what my father said, the man never seemed persuaded that he was safe sharing the room with us. I could feel it.

I was starting to understand how expansive my father's peaceful spirit was. Still, I found myself getting agitated watching him always being the first one to reach out his hand as a friend. I studied how he engaged with the boy's father who put no effort into making a connection with him.

My father was the one who had to leave his house and cross checkpoints to get to see me, not him. My father was the one who had his house occupied and had watched his son get shot, not him. Yet, my father was the cheerful one, not him. In fact, the man was aggressive and pessimistic. If my father said anything from a Palestinian point of view that did not sound agreeable to him, the man would withdraw. I found it sad that my father's optimism upset the man's paranoia about the future.

Mohammad was picking up some good Hebrew. He could make sense of what was being said around him. He enjoyed that. He told me that he heard the parents of my roommate complain about having to share the room with a Palestinian. That bothered me, but I decided not to be upset about it. Also, it helped when I heard that the nurses were telling the father, who was a settler, to get to know me before he judged me.

I appreciated that the nurses saw me as an individual. Sometimes, however, I had to admit that, even though I was treated very well, I felt uncomfortable being there. I had told the staff and others at the hospital that I was not "throwing stones at the soldiers," which had been their presumption when I first arrived. They were very honest about it, for which I was grateful. It allowed us to talk more freely.

My father came to visit several times on weekends. With the help of the German ambassador and Rudolf Walther and the pressure they applied on Israel, he managed to get permits for my younger siblings to come with him. On each visit he brought one of them. I loved seeing them. Misoon and Mohammed Salah would take turns pushing my wheelchair and telling me about what they had seen on the way to the hospital. I enjoyed watching them have fun playing in the children's games area. Zana was always looking at me with her gorgeous eyes as if I were from outer space.

My father always brought Seema some Arabic sweets like baklava from Deir. She loved them.

"It is what I grew up with," she said one day when he brought something special. "We had a good life in Iraq, in a lot of ways. People outside only hear about the bad things." That sounded a lot like Palestine to me. I wanted to ask her more, but I never did.

Eventually, my mother went home with my father after one of his visits. It had been hard for my parents to be away from home and from each other. Also, even though we had help, getting permits to leave Gaza was always a nuisance.

My mother said goodbye to all the staff and many of the parents on the floor. She had become a part of life in the Sheekum, always helping where she could and spreading good feelings wherever she went. The plan was that my parents would return and see me again soon.

"By then, you will be walking," my father said.

12

The Miracle

I did not mind being alone, because as I got stronger and the wound began to heal, I enjoyed having some space of my own away from my parents. The hospital kept us busy. Volunteers came and played cards with us. Some of them were university students who were disabled. Others brought toys, CDs, or sweets.

Once a group of Hassidim musicians dressed in their big black hats and long beards came during Passover and started singing and dancing around my bed. One had a guitar, another a harmonica. They were singing Passover songs like *Dayenu*.

"Day Dayenu, Day Dayenu, Day Dayenu, Dayenu, Dayenu."

I liked it. After a couple of songs, they asked me where I was from. I told them bluntly, "I'm from Gaza." They looked at each other, then left without saying anything. I do not think they regretted singing for me, or so I hoped.

It was fun to roll around the hospital and to look at the beautiful countryside from my window. Everything was beautiful—beautiful buildings, beautiful roads, beautiful trees, and beautiful people. I would go out as far as the main entrance of the hospital's compound. I looked across the fences and watched the tops of buses appearing and disappearing. I observed the people walking on the streets and the cars beeping to one another. I wanted to go outside, but I knew that I was free to move around only inside the hospital.

As I observed the people on the streets, I thought about all the stories I had heard as a child—how this same land had once been part of Palestine. As much as I was now living with a newfound understanding of the Israeli people, I was struggling to incorporate the story of historical Palestine into what I was seeing and experiencing.

"My own people used to live here," I told myself.

It was less than a day's journey on camel or by boat from Deir el-Balah to old Jaffa next to Tel Aviv. For centuries, my ancestors must have gone back and forth from there to here countless times, as well as to Jerusalem and all the other towns now off limits.

I thought about all the lives that were destroyed by Al-Nakba—the Catastrophe of 1948—and all the Palestinian families who had been forced into exile until further notice. How could this good and beautiful place that looked like it had dropped from heaven fail to allow all people to live in it? Even when I tried to let go of my thoughts, I found myself haunted by the war-torn images I had from the Strip. I knew that war was still going on at home, and that the Israeli army was still occupying my house. None of that made any sense to me.

There was a hill outside my window. Often in the evening I would sit in my wheelchair and watch the sun go down behind it. Sometimes I would even sing the songs of the late great Egyptian composer Mohammed Abdel Wahab. His buoyant words would elevate my soul:

> *Love is our soul . . . our land and sky;*
> *Sadness . . . tell me where will you come from;*
> *You are the eye of my heart . . . the most beautiful eyes.*
> *Let us live . . . enough speculations;*
> *Let us alter our life into one night*
> *Away from deprivation and sadness.*
> *If anyone blames us . . . let us say*
> *Without love . . . there will be no one on Earth.*

Sometimes I would hold my legs and talk to them. I thought that if I did they might listen and get stronger. Sometimes my tears fell without my permission. Sometimes I would just keep on sighing and sighing until my mouth got too dry and my tears flew out the window.

In the midst of the pain, though, I became aware that a miracle was unfolding within me, not only in my body but also in my soul. Being in this hospital was helping me begin to understand what my father had tried so hard to teach me. One day, while talking to him on the phone, I told him, "One Israeli soldier shot me, but there are many Israelis here who are working hard to save me." I was starting to understand what he meant about "peaceful coexistence."

He responded, "Think of this as the opportunity that can change your life." Then, sounding a bit like an alchemist, he added, "A new door has just opened for you, Yousef. Do not be upset. Do not be upset."

* * *

The day when he took my mother home from the hospital, his parting words to me were, "Yousef, if you want to learn to walk more quickly, ask for more time in physical therapy. Some of the best therapists in the world are in this hospital. Let them help you."

I knew he was right and I loved working in rehab. The therapists wanted me to kick a ball, or try to take the stairs on a machine they brought, or walk against the wall then try to walk without using the wall. Sometimes I fell, and sometimes I went longer before falling again. All they said was "good job," "Bravo, Yousef," or "Wow! Nice work," and similar things. I was dedicated during physical therapy. Although I usually liked to joke and have fun, I did none of that in therapy. I was pretty straightforward with the therapists and they with me. They said, "Do this," and I did it as best I could. All I wanted was to walk again, desperately.

The first time I tried to get out of my wheelchair, I collapsed. Still, I got over the fear of falling very quickly. I knew that it was going to happen. Every time I attempted to rise from my chair I would feel a sense of excitement and hope. With the trainer, a beautiful woman, waiting for me to rise, all I wanted was to impress her and get up like a man. Despite the nagging pain, I was enjoying the process.

As I worked, the most common question I was asked over and over again was, "Yousef, can you try and lift your legs?" Of course, no matter how hard I tried I never could. The pain was merciless. With time, though, my legs began to obey my orders again. Physical therapy taught me that I still controlled my brain and that by doing so I would control my legs again in time.

So, slowly, my legs started moving again. Just a little, but they were moving. I felt that my body was like a machine that needed oil to move smoothly. I could almost hear my leg bones crunch and make noise. I had not moved them for so many months. First my left leg recovered, and I could move it. After a while, my right leg began to move also. Everything, bit by bit.

One day the trainer encouraged me to kick the ball as hard as I could. She said, "You can break things if you want, so don't hold yourself back. Just kick the ball." When I did, I kicked it right into her chest, and for a moment she was struggling to breathe. I kept saying, "I'm sorry." All she did was laugh and say, "Good job. You did it. We just need to teach you how to aim next time."

A few days later she told me, "I am going to take your chair away. You don't need it." She gave me a cane. I did not know what to do with it. "Why not stand up and use it?" She smiled.

I wanted to rise on my own, but I could not. I sat there for a few minutes telling myself, "If I am ever going to walk again, I have to do this." Slowly, I leaned forward and without using the arms on my chair I raised myself up. I was standing. I was standing!

The therapist shouted, "*Sababa* [Bravo]!" and slammed her palms against one another excitedly. I wanted to make sure I had really done it on my own. I sat back down and got up again. I did that about three times to make sure I had not been hallucinating after all the drugs they had given me. I looked at the nurse. My eyes were tearing up. She hugged me. It was my first moment of independence, the first thing I had done on my own since I had been shot.

I wanted my father to be there, to see that I was a strong person who had persevered. I was not just someone who would disappear because of a bullet in his back. I knew he would be so proud of me if he could see me walk again. Everyone made a big fuss and came and congratulated me. Of course, after half an hour I was treated like everybody else again.

Now they wanted me to try walking. I held on to the cane, and was told to walk and try not to fall, which I did a few times. Then, one morning I started to walk again. At first, I had to lean against the wall, but then I did not. I no longer needed to sit in the wheelchair.

I again imagined myself as the good guy who gets knocked down, then builds himself up, and comes back and wins. That helped to keep me going. The day the doctors declared me able to walk, I celebrated nonstop. I just laughed and laughed and laughed. I wiped my tears away, hoping that they were not seen by the trainer. Dr. Brezner and the trainer were amazed and clapped their hands, "Wow, Yousef! Wow!"

The bandages on the wound still needed to be changed, but by then I had become very good at doing that myself. I said, "One, two, three . . ." And when "three" came out of my lips I peeled the bandage no matter what was going on in the world. The faster I did it, the less pain I felt.

Three bullet fragments had lodged beside my spine and had damaged the nerves surrounding it, but not the spine itself. So maybe I could say that luck was on my side. Being shot while UN staff were present, with a car to take me immediately to the hospital, made a big difference. Plus, my father and his German friends had managed to have me admitted to a fine Israeli hospital, where I had received extraordinary care.

The positive attitude and love of my father had surrounded me while I was at my worst and had gradually seeped into my being. My body had needed to heal and rebuild its damaged tissues, but it had all the support required to do so. I felt so empowered and transformed. Most importantly, my spirit was enriched by a new set of values and dreams.

The real test for me would be going home again. Most of the people I knew in Deir had thought I would never walk again. They would be shocked when they saw me.

My parents were given permits to return to Tel Aviv to pick me up and take me home. When they saw me standing and then walking, they were in total joy. My father said to my mother in front of me, "If he puts his mind to something, he just does it!!!" They hugged and kissed me and I was happy to let them do it. I loved rubbing my face against my father's face again. It had been too long, and I had missed that. My mother kissed me like the old ladies, my grandmother's friends, did. I had to rub my cheeks after every kiss, and there were many of them.

She acted annoyed and said, "I made you." I kept rubbing my cheeks anyway and said, "I don't care." We all laughed.

My father had brought a tray of Arabic sweets like baklava for every single nurse and all the doctors who had worked with me.

There were so many people I wanted to say goodbye to. I wanted to go home, but I was sad to leave them. I knew that I was really going to miss Micha and the nurses. I had been there for what had sometimes felt like forever. I had a couch in the hallway where I liked to sleep. I had the shawarma joint. I had the escalator competitions. I had FIFA. UNO games. I even had a teacher who attempted to keep me up with my missed schoolwork. I had made a life there.

It was extremely hard to leave the other patients who had been in the hospital before I had come and who were going to be there a lot longer. It made me feel embarrassed knowing that they were way more badly injured than I had been, and in worse shape than I had realized.

I looked for Seema, but could not find her. She was off that day, which upset me. I knew, though, that I would have to come back for checkups, so I figured I would just wait until then to say goodbye. She would laugh at that, because that was how she was.

Outside the hospital, a Palestinian ambulance was waiting. It had brought somebody to the hospital and was heading back to Gaza empty. The driver was looking for passengers. If we traveled in it, it would be express to Erez. The driver demanded fifty dollars. My mother hesitated. My father agreed to pay it. If we had tried to get a taxi, we would not have found one. Taxis do not want to go to the border area, and we would have had to take a bus.

As we all got in, I knew that my days of being special were about to end.

* * *

We left Israel through the border at Erez and had the usual long and tiring wait. Then we went through the tunnels and past the faceless guards and crossed over. We got a taxi on the other side. I was riding alongside my father and mother like before. As we drove home, everything in the Strip looked the same, only with more destroyed buildings. I learned that a few boys I knew had been killed by the Israeli army. My parents had not wanted to tell me before.

I knew that the soldiers were still in our house. I had no idea how they would react to me. I knew, though, that I had to avoid bringing my anger into the house with me.

I wanted to look at the house, the land, the base, and everyone there with new eyes, the eyes I had developed while I was away. My head was filled with the memories of how all those things had been when my home was a paradise. Maybe I thought that everything would be like that again, that the same kind of miracle that had happened to me would have happened to it.

When we got home, however, the land around our house was more like a yellow desert than a farm. The Israelis' Merkava tanks and bulldozers had crushed the sand so finely that I could now breathe it with the air. I could not feel any connection to those images from my past that I had stored forever in my brain.

My father had been saying to me for some time, "Do not be angry with the soldier who shot you. Do not wish him evil. Challenge him to ask himself why he shot you, why he shot a young boy who did not and does not wish him evil."

As I walked through the door of the house for the first time since I had left it in the UN vehicle, the words of my father played like lyrics in my head. It was as though I were hearing a magnificent symphony. His message was sinking into me. Still, when the ever-watching soldiers looked at me like nothing had happened, that upset me.

I knew that my father's hope had played a major role in my healing. His positive attitude and his optimism had inspired me and helped me to see what was possible. I was starting to see the world and the people in it as one connected whole. I was beginning to think there might be a meaning to my life and to wonder how I could learn what it was.

When we sat down on the floor in the living room to talk or play board games like Monopoly in the shadow of the soldier guarding the living room door, I kept looking at my parents as they stretched out next to each other and my siblings. I had missed them so much and was overcome with love for them, our land, and our people. I felt as though, for the first time, I was beginning to understand what it meant to truly love and to truly love being a Palestinian.

At the same time, I knew I had fallen in love with many of the people I had met in Israel. Angels could not have treated me more kindly than those Israeli nurses in Tel Aviv. They had smiled on my frightened face, played with my hair, seen me naked more than my own mother, and watched me in tears of pain. They had saved my life with their caring. This love would not have been possible without my profound love for my father and his values.

I wondered why the Israeli people felt they had to send the soldiers to my house and why I had been shot. I wondered why everyone did not feel the love I was now feeling. I understood what my father meant when he said of the soldiers, "They are just children, forgive them."

To discover my own humanity and to know that my apparent enemies were also human was my father's most important gift to me. This changed my life and allowed me to forgive and to dream of a future when all of this had become history, and we would be living in true harmony with the Jews—just like before. This gift would set the course for my future journey.

Part Four

THE JOURNEY

"Never let hatred for any people lead you to
deviate from being just to them.
Be just, for that is closer to being mindful of God."
—The Holy Quran, Surat Al-Ma'ida (5:10)

13

Escape

The circle of the wound was gradually getting smaller. It was comforting to see that, day by day, it was disappearing and burying with it the bullet fragments still in my back.

It was strange to be home. We were still "in prison," and I was not happy about that. Much of what I had learned while away, and much of what I now wanted to share with others, seemed pointless in those surroundings. My vision of what could be was challenged by the noises of war all day and the constant disruption by the soldiers.

It seemed crazy to have gone through such a life-changing experience and come back to find that nothing had changed. Nothing. Some days I felt useless and worthless. I knew, though, that no matter what happened, I must never allow myself to think like that. To do so was admitting defeat.

I am not worthless and I have a voice. If that voice is suppressed before it is even heard by the world, who will decide whether it is good or bad?

Now, more than ever, I wanted to leave Gaza. The palm trees that had stood tall for so many years had been demolished by the bulldozers. Even the grape vineyards that had stretched for centuries along the coast were vanishing, one after the other. We used to park the car on the highway and get a box of grapes freshly picked from the vines. The good ones grew really close to the beach. What harm had they caused anyone?

I pondered endlessly about going to school in another country, but I knew I did not want to be an engineer or a doctor like my older brothers. I tried to keep these thoughts to myself, but it was like living life with a plastic bag strapped tightly around my face. I kept waiting for the moment when I could rip it off and finally allow oxygen into my lungs.

One day shortly after I came home from the hospital, a bulldozer driven by a soldier began knocking down everything in front of our house. It was about to take out the phone lines. Other bulldozers had already done this a number of times, and we would have to wait for days until we managed to find a technician brave enough to come near our house and restore the line.

I was sitting on the veranda when I saw the bulldozer approach. I still had to walk very purposefully. No sudden turns. I had to be conscious of how and where, with every step, I placed each foot. The bulldozer, though, wiped out all those instructions from my mind.

I walked down the driveway as fast as I could to the street. There I just stood in front of the bulldozer's sharp and shiny claw with my arms raised up high. It stopped a couple of feet in front of me.

"Take another route," I shouted at the soldier driver. "We've just fixed the phone lines. There is no reason to knock them down again." He sat in the cab looking at me, trying to decide what to do. I had already decided what I was going to do: I was just going to stand there.

I was about to win the duel, but my grandmother Zana came running down the driveway toward me shouting and screaming like the sky had fallen to earth. As always, she was dressed in her brightly flowered blouse, long black skirt and long white headscarf wrapped around her neck. Usually just a glimpse of her generated good feelings. Not at that moment.

The soldier saw her shouting at me to come home. He opened the window of the bulldozer, stuck out his gun, and fired a few shots to the air. I was so upset with my grandmother for interfering with my battle with the soldier that I unleashed all my fury onto her. I do not want to write down the words I shouted at her. I used a tone I had never heard from myself before. Instantly I was sorry. By then her eyes were already filled with tears.

I put my arm around my grandmother's shoulders and pulled

her to me. The soldier kept firing, this time because he knew he had won. Without looking at him again, I slowly walked with my grandmother back to the house. As we went I heard the telephone poles behind me falling with earth-shaking thuds.

The phone lines were cut off again. Eventually, my father got them fixed. In the meantime, the way I had treated my grandmother was more painful to me than the wound in my back.

She was becoming slower and slower every day. My father was already telling us that he had a feeling she was about to leave us, but I thought that was just something he was saying.

One night we were awakened by the sound of soldiers coming through the kitchen door and the two front entrances. We had not thought they would come, so we were sleeping in our bedrooms. Instead of ordering my parents to move us to the living room, the soldiers pounded on our doors and pulled us out of our beds. This time a different soldier handled each of us individually rather than ordering my parents to gather all of us and move to the living room. It was so strange to see the way the soldiers transformed themselves into human robots in situations like this.

They walked us outside and herded us into an open area between the back of the house and the base. A military jeep pulled up next to us. We were surrounded by soldiers. It was dark and cold. I was looking for somewhere soft to stand, because I was barefoot. I silently regretted not following my mother's rules about not walking around barefoot. I saw a tank lowering its gun, and adjusting its aim at the back of the house.

My grandmother was still inside, and one of the soldiers asked my father to bring her out. When he finally did, she said she could not stand up. The captain in charge wore glasses and looked as if he was out to save Private Ryan. He made my grandmother walk to join the rest of us.

"They are up to no good," she told my father. I looked at the house and the tank and wondered about possibilities that had never occurred to me before. *What are they going to do this time?*

The captain stood in front of my father and, after adjusting his gear, asked, "Do you have a terrorist in the house?"

"Absolutely not," my father replied calmly, but decisively. He looked surprised to be asked such a question in the middle of the night. He must have thought about all the negative possibilities in an instant, and stated, "We are committed to peace. We don't support violence."

"OK, you go upstairs, you search and you come back," the captain ordered my father. "If there is no terrorist, you come back alone. If there is, you come back with the terrorist, OK?" My father quickly took off back into the house. Many terrifying ideas crept into my head as fear invaded my soul. What if there was somebody in the house that we did not know about? But how could this be? The soldiers were living upstairs.

On occasion, some Palestinian fighters would come around our area and shoot at the base and at the soldiers. Maybe one of them had managed to get into the house. My father thought shooting at the soldiers was pointless, although the soldiers certainly seemed concerned about people shooting at them. My father even suspected that some of those firing the shots had been hired secretly by the Israeli army. Sometimes these "fighters" would shoot no more than twice. Then a bulldozer would suddenly appear and knock down some buildings or destroy some family's farm. My father believed that that was what happened the night they demolished our greenhouses.

One of the soldiers guarding us eased my worries by offering a seat to my grandmother. As soon as he had, the captain yelled at him.

"*Ma ata ose* [What are you doing]?" the captain screamed. It was easy to see what the soldier had done, so why the question and why the anger in it? I had never seen the soldiers yell at one another before. I watched, with a great sense of interest, to see what would unfold. This soldier must have been high up, otherwise he would not have stood up to the captain.

They shouted at each other in Hebrew for a few moments. When they finished, the captain strode off and the soldier helped my grandmother get seated in the back of their jeep. By then, however, she wanted to stand, because she was worried about my father who had just made it to the top of the house. I could see him climbing up the stairs. When he reached the last few steps, something exploded all around him. It was too bright to look at, and loud. We all dropped instantly to the ground, while keeping our father in sight. It was like he was living in his own separate scene, while we lived in ours.

They spoke to him through their megaphones and ordered him to remove his shirt. When he came back he said to us, "I don't know what they are doing. There is no one upstairs."

After some more waiting, the captain told us to go inside the house and ordered my father to get into the back of the jeep. We all paused, refusing to move without my father. I got up in the captain's face, grabbed his gear and tried to stop him. For a moment, the other soldiers just stood by and watched. In a matter of seconds, though, things escalated, with my mother and grandmother pulling me and the soldiers pulling the captain, trying to split us up. When finally we were separated, the captain adjusted his gear then pointed his gun at me and shouted, "Boy, I shoot you right here." My instant reaction was to turn around and give him my back. To me he was a coward, and by giving him my back I wanted to reaffirm that.

The captain casually stated to my father that it had just been a drill. "We need to be prepared," he said. "There will be more."

My father slapped his palms against one another and looked at us in a way that said he had already figured out it was just a prank. I decided the captain was probably trying to get promoted.

Perhaps the soldiers thought that "drills" like this would finally drive my father out of the house. He was never going to go, but I was ready to go almost anywhere else. I wanted to leave and find another place where I could have a normal life. I had been studying the world map and dreaming of the places I wanted to see: the Golden Gate Bridge, Big Ben, the Eiffel Tower, the Great Wall, and, of course, the Santiago Bernabéu Stadium in Madrid to watch my favorite team.

If anything, the urge to leave and see other places only got stronger after moving back home. At Tel Hashomer Hospital, I had a glimpse of things beyond Gaza. I now knew that there was a whole world waiting for me on the other side of the sea. Like a bird, I was eager to try my wings and escape my nest in search of something better.

* * *

In spite of all the love I had experienced from the nurses and the doctors while I had been away, and all the love I could feel now, it was very hard to face the soldiers every day and not get angry. I worked hard to forgive them and to be a better person, but they continued to do only harm. It was agonizing to hear them order my father around and treat him like a prisoner in his own house. Yet my father continued to treat them as though they were honorable human beings.

Sometimes, strangely enough, a soldier would be polite enough to ask, "Do you mind if I smoke a cigarette indoors?" My father would kindly answer, "Go ahead, feel at home." They were to feel at home? This was now their home? That seemed like total insanity to me. The whole thing had to be one big joke. How does one learn to be polite despite the impoliteness of others?

I knew, however, that I had to keep such feelings to myself. How could I complain about my situation when my parents' situation was even worse? Having watched them work so hard to help me survive, I had to find a way to repay them. I came to believe the only way I could do that was to find a way to leave. I could see what a toll everything was taking on my father, fighting for what he believed in and raising his children at the same time. He never sat still and I doubt he ever truly relaxed. I watched his hair turn whiter with each new day.

The news of Yasser Arafat's abrupt death on 11 November 2004 brought all the Palestinians together like never before. Everyone was sad, even those who had opposed him at times. We could almost live with his public humiliation as our leader by the Israelis, since the whole nation was living in a prison, but to have him die was the hardest thing for all the children of the Intifada. I knew how great he was by the amount of unity I saw flood the streets of the Strip as news of his death spread.

A couple of days later just before dawn, my grandmother asked me for some water as we were preparing to fast for another day of yet another Ramadan. Sometimes I would pretend that I had not heard her requests, hoping that someone else would do it or she would forget. That day, though, it was just me in the living room with her. She asked me again and grudgingly, as only a teenager can, I went to get it. When I came back just moments later, I saw my father on his knees calling to her, but he got no answer.

"Grandmother has passed," he said quietly. I was stunned. For all the shooting and killing that had gone on around us, this was the first time a death had occurred in our house. Death had been casting its deep shadow over us for some time—since the first time the settlers had attacked us, since the soldiers had started shooting at us, since my father had been injured, since my brother had been shot as he put out the fire, since I had been shot in the back. Yet death had always failed to actually take one of us with it. It now prevailed. To take my old and weak grandmother was to me a cowardly victory.

We all gathered around her, but I was especially devastated. I remembered all the times she had said, "No, Yousef is very kind, he always helps me." I kept adjusting the pillows under her head with such care and sadness. Just when I had thought that I had survived death, I was reminded that it is always near us. At some point, it will take me.

We carried her body and placed her gently in the car. I drove with my father to the hospital, along with Zaid and Mohammad Salah. He wanted to make sure that she was really gone. When I was younger, my mother had told me the story of how he had refused to believe that his father had died.

"He insisted on taking him to the hospital," she said with a faraway voice. "I thought he was just reacting to the shock of his father's death, but it was a smart move. The doctors determined he was only in a coma, and he lived for another year." This time it was a different story.

We placed her lifeless body on a bed in the emergency room and waited for the doctor to come.

"My condolences," the doctor told my father as he took off his stethoscope. My father nodded, shook the doctor's hand, and told us to carry her back to the car.

On the way home, I held on to her body as tightly as I could, feeling so sad she had left us. She was so light. I rested my face on her and wept deeply. When we arrived home my father jumped out of the car to go find the soldiers. He walked slowly toward the tower, knowing that at any minute he might be shot.

"Hello. Hello," he called to the soldiers in the tower. No one answered. It seemed like no one was there, although we knew they had to be.

"Hello," he called a few more times. Finally, a voice over a loud-speaker told him to return to the house.

"My mother has just died," he called up to them. "I want to have the funeral here."

"You are not allowed to bring anyone here," they told him. "You know that."

"My mother has just died," he repeated. "This is her home."

"Use a neighbor's home."

"The women need to come and prepare her body for burial," he went on, "and . . ."

The voice from the loudspeaker cut him off. "Return to the house immediately."

My father looked like he wanted to say something more. Instead he turned and walked back to where we were waiting for him.

We carried my grandmother into the house. My mother gently washed her body by herself, and then wrapped her in white sheets. We recited funeral prayers at Abu-Salim Mosque. Later, my father and my brother took her to the cemetery in Deir el-Balah to lay her next to my grandfather's grave. I was so sad, I just watched from the far side of the cemetery.

We held the funeral at the home of our neighbor, Abu Immad. Up until the last minute, I kept thinking that the soldiers would send a signal to my father that they had changed their minds and we could gather in our own house, but they never did.

At the funeral, people told many stories about my grandmother. I learned a lot that I had never known. Her mother had died when she was giving birth to her, so her father abandoned her and did not treat her well. He remarried and made her serve his other children as a maid; her stepmother did not treat her well either. She had inherited land from her mother because she was the only child, but when her father died she got nothing from him. She married my grandfather and had two sons and two daughters. After my grandfather died, my father took her in and she lived with us in our big house.

It was very sad to hear her story, but it made me feel even more proud of my father. I was so grateful that he had provided her with a good life. We all loved her so very much. Having her live with us had helped to teach me the importance of being kind to others.

* * *

It was not long after my grandmother's death that my father was diagnosed with diabetes. It was clear that the stress had become too much for him, even though he tried not to show it. I began to worry about him.

About that time, I overheard him talking with someone about "Seeds of Peace," a leadership development organization that works for peace. The program brings young leaders from conflict zones around the world to a summer camp at a place in America called Maine so that they might gain a better understanding of their enemies.

"Amreeka?" A bomb of curiosity had exploded in my brain. "Can I go?"

My father had always pushed me to learn English. He had tried to get me to read English dictionaries, English novels, and whatever else he thought would make me an educated man. He maintained that the only way for a Palestinian—rich or poor—to make a decent living was to become a doctor or an engineer, even though he himself was a teacher. He would say over and over, "The land will eventually be split among the eight of you and it won't be enough to take care of your children. You must study important things so you can take care of yourself and your families." He always said that it did not matter how much land you had, an unjust army could come and confiscate it—but they can never take your education.

I did not disagree with him, but I did not want to be either a doctor or an engineer. Why, just because I was a Palestinian, did I have to go into a particular profession? Why could I not become anything I wanted to be? Why live a life of limits?

The one thing I knew for sure was that I wanted to do something important and to make my father proud. Day by day, my father's message was becoming my own. By now, our relationship had grown to be more than father and son or teacher and student. We were beginning to relate as one human being to another. When we spoke about his beliefs, it was as if we were two adults talking about important ideas. When I was in a fight at school, he would still reprimand me as before, but later he would talk to me about the importance of treating everyone respectfully. He wanted me to understand that to be just, one must be fair.

We would spend hours talking, and sometimes he even seemed impressed by what I had to say. When we were done, it would be back to the same old, "Do your homework, Yousef. Stop wasting time." I hated that part. After all the interesting stuff we had just shared, he still treated me like a teenager. I really hated it.

I started to get really excited about the Seeds of Peace idea. I kept asking him, "What is this Seeds of Peace thing?" Even though he had told me many times already, I would ask again. I did not want to risk him forgetting about it. I was desperately looking for some way to get away, and I sensed it had to happen before my final year of secondary school. For some reason, I felt that if I did not leave before then, I would be doomed. If I did not live up to my father's dreams for me, he would be devastated. I feared I could end up in Germany like my brothers, and I certainly did not like that idea. I had heard from them that in Germany a foreign student had to leave the country if he failed one exam. One single exam and you were out! I did not want to have to live under that kind of pressure. I had enough where I was.

I kept nagging my mother to remind my father about Seeds of Peace, and to get him to sign me up. I knew that he gave his full attention to what she said when they went to bed at night. I was even more attracted to Seeds of Peace when I found out they also had a program that helped participants apply to boarding schools and universities in the US.

It all made me think of the letter I once wrote to Real Madrid, inviting the club to come "discover my talents." Whenever I wrote, I knew it was unrealistic. Boarding school, though, made perfect sense. I was determined to go to this amazing summer camp. There I would shine and get myself into a boarding school in America and avoid having to come home to finish secondary school. Boarding school in America would solve all my problems. It was my one and only dream. It was the golden key, I thought.

I knew my father would not oppose this idea because he had almost sent my oldest sister to Italy, after she had won a scholarship to go to a UN boarding school. Instead, he offered to send her to a university in Germany to join my brothers when she finished secondary school. She got a ninety-six percent and went to Germany to live and study with my brothers. I envied them for being out, for being free, for being able to create their futures. I had to find a way for myself or it might be too late.

I became obsessed with the idea of going to America and was convinced that it was the right thing to do. I began reading about Seeds of Peace and started speaking English around the house. My father had always said, "If you know English, it will open any door you want to enter." I had never paid much attention to that, but now I could see that English was essential. I was willing to do whatever it took to get out and, if it required learning English, then I would learn English.

At first, Seeds of Peace did not want to take me. They said I was too old. The age limit was fifteen, and I was already sixteen. I was so frustrated. It seemed to me that no matter how hard I tried to create my own opportunities, new obstacles would suddenly appear out of nowhere. I did not give up. I kept bugging my father, begging him to do what he could to ensure I was part of the Palestinian delegation.

I kept asking my father why the US would create an organization to facilitate Middle Eastern peace when they refused to make Israel live peacefully with its neighbors. Filling his teacup, he would respond thoughtfully, "Every country has its good and its bad. Perhaps they recognize that the world needs to be safe for all people, but don't know how to achieve it."

I was beginning to realize how complicated this peace thing was going to be. I had been shot by an M-16 bullet, which had been made in the United States, and yet I wanted to go to the United States to promote peace. It was a puzzle, but by then I already understood that life was a puzzle.

In the end, I was selected to attend the Seeds of Peace International Camp as part of the Palestinian delegation. I suspect the decision was partially due to my having Mr. Khalil Bashir for a father and partially because I was a Palestinian who had been shot by an Israeli.

I wrote down some ideas I wanted to share and even wrote a few speeches in case I had to give one. I felt so involved and committed. I had already had plenty of experience with all the journalists who had come to my house. I was no longer scared of being in front of a camera or of speaking in English with everyone in my family staring at me, though previously this had made me feel self-conscious.

* * *

Getting to Maine was the next challenge. It had always been hard to get permission to get out of the Strip, but now Israeli checkpoints inside the Strip were making travel even there a nightmare. At one Israeli checkpoint near my house, the line of cars waiting to get through often stretched out of sight. They could be stopped there for hours, even days. Little children would appear out of nowhere to sell cups of hot tea or cold carob syrup to people stuck in the sizzling heat. Traveling to a town only a mile away could be a big hassle, and only those who really needed to travel did so.

Luckily for me and the other twelve Palestinian students in the Gaza delegation, government officials and international personnel were making all the travel arrangements which, of course, made me feel very important, like I was a Real Madrid player.

When the day came to leave the house and join the rest of the Seeds of Peace team, my family gathered in the living room. I hugged them all. My father was standing at the door.

"Yousef, remember to be kind to everyone you meet. Laugh, be happy, and, most importantly, spread the message of peace."

I was always uncomfortable when he talked to me like that, and he knew it, but he would say such things anyway. I had to do well in Maine. I knew the two Palestinian teachers who were traveling with us, a man and a woman, would report to him every detail of what I said and did. Besides, I was determined to make the most of the experience.

As I got on the bus, I was aware that I was on a very important journey—or so I told myself. In my mind, it was all about my inner motivation and understanding of things around me. I found myself thinking a lot about my father, and how he had managed to become a man of such limitless decency. The further I found myself away from Deir the prouder I felt of my father, and I thanked God for him.

There was a time when I had thought of my father as a weak man, because he was always forgiving others. He had claimed that the only revenge he needed was to see his children successfully living on his land.

Now, in my mind, I played a slideshow of images of him being kind to his family. How he had been so loyal to his two sisters, insisting on their receiving their fair portion of their father's land. How he had passionately cared for his mother, until her last moments. She had passed peacefully in the knowledge that her son loved her. Something in me kept arguing that he was too kind for his own good, but another part of me was confident that I would excel if I could completely embrace his values and ideals.

Slowly, our bus inched closer to the Egyptian border.

14

Seeds of Peace

By the time we arrived in Cairo, I felt warmly welcomed into her arms. We had crossed the border at Rafah, which took forever, and I entered the land of Egypt for the first time. The military men at the border had hilarious accents and AK-47s that looked bigger than they did. They moved around us in their white uniforms from one bus to another.

We stayed at the Hilton hotel in the middle of Cairo. I had never stayed in a real hotel before. For my uncle's wedding, I had been to one by the beach just outside Gaza City that was supposed to be the most luxurious hotel in Gaza. The Cairo Hilton was in a totally different class.

The other students in the group all knew each other and did not seem to want to interact with me. I was the only one from Deir. I just kept to myself. Omar and Hossam would have been shocked to see me so silent, instead of plotting some prank.

When one of the teachers traveling with us announced who would sleep in which room, I saw that the others had all hooked up with their friends. I did not mind; I was too busy relishing the rhythm of Egypt, inhaling its essence. I was out of Deir! I was out of Gaza! I slept like I had not slept in years. I put down all my shields and trusted that in the morning everything was going to be fine. I kept on picturing myself at camp, walking around and making new friends. Laughing and finding my new opportunity.

In the morning, the hotel buffet looked exactly like what I had seen in Egyptian movies. The food was beautifully garnished and each chef was dressed like those I had seen on the Egyptian cooking shows during Ramadan. It was magic becoming real. I so wished my mother were there with me, if only for half a minute, to see how real it was.

If I had paid more attention to all the instructions being given, I would have learned when we would be flying and where we would be landing and what was going to happen, but I was far too excited to listen. I just got on the plane ready to sit there for as long as it took to reach America.

After endless hours, the pilot finally announced that we were in American airspace. I silently cheered, "Wohhooooowwww!!! Amreeka!"

Time suddenly turned heavy when the plane landed in Boston. I was impatient to see everything immediately. Instead, I had to show my passport and answer all the usual questions. It took a couple of hours for the whole group to get through customs and out of the terminal.

When I finally saw the streets and highways almost hanging in the sky, I knew I was in a different universe. I knew it was still Planet Earth, but it was so hard to believe I was now in the United States, on the far side of the ocean.

Eva from the Seeds of Peace office was waiting to welcome us and take us to camp. After boarding a large air-conditioned bus, we cruised through Boston. This city looked magnificent and I was instantly seduced. The rivers and the trees. The cars bigger and the streets wider than any we had at home. It all made me feel pity for the peoples of the Holy Land. While they were busy fighting, people in other places were hanging bridges in the skies and stuffing tunnels under rivers. Why can people not live in peace, enjoying the luxuries of their ingenuity instead of living on the thorns of their tragedies?

It took us almost three more hours to reach Otisfield, Maine, and find our camp hidden in those surely God-made forests. The whole drive there seemed like a dream. I stared out the window and attempted to take it all in.

When we arrived at camp, many people with cameras took pictures of us as we got off the bus. I soon discovered that our Gaza delegation had arrived three days after the opening day. There had been some delays in getting our visas and the fiasco of getting through the Israeli checkpoints. I was upset that I had missed three days already.

After I got my stuff, I stood there trying to absorb it all. There was so much going on.

"Welcome, Yousef. I'm Adam." A friendly young man greeted me with a handshake before rushing to pick up my luggage. "I am going to be your counselor while you are here." He did not seem much older than me, though he was a university student. He kept telling me how happy he was to meet me and trying to help me feel welcome. I was feeling very special until I overheard the other counselors saying exactly the same things to their arrivals. I was definitely not the only "special kid" around.

In fact, I sensed I was really going to have to prove myself if I wanted others to listen to what I had to say. They all seemed to have a special thing about them. They all made me feel as if I were interacting with some future stars, good stars. Everyone looked cool and walked around like they were on a mission.

Adam took me to bunkhouse 13. I was asleep the moment he showed me my bed, but deep in the middle of the night I opened my eyes and listened to the breezes as they blew all around me. I waited eagerly for morning to meet everyone else in my bunkhouse. I had to make every moment count. I knew I was about to make lifelong friendships.

When the sun finally shone upon the world again, I heard the strong and loud voice of a man. His tone was that of an army chief at a military camp. He allowed no room for hesitation. After the ringing of a bell he shouted, "Liiiiiiiiiiiiiine up! Liiiine Up! Liiiii-iiiiiiiiine up!" That was Tim, the head of the camp. The big boss. I learned later that he was a legendary secondary school football coach.

As time passed, voices from other bunks began to rise as windows and doors were slammed. Everything was made of wood. We were right out in the wild, in a very green and inspiring forest.

As I looked around, I saw that a Seeds of Peace T-shirt and sweatshirt had been folded and placed at the corner of my bed. Details like that immediately made me happy and filled me with a sense of responsibility. I thought of how football players have their shirts folded and laid out for them before a match. They have more serious things to take care of than worrying about finding their uniforms and gear.

A gentle voice approached me as I smiled at my new T-shirt. Daniel, the second counselor in the bunkhouse, introduced himself to me. I soon learned that Adam and Daniel would become my bosses. Daniel had a guitar hanging by his bed and seemed like a peace-loving person. He led the Sunday service for the Christian Seeds.

For all their friendliness, none of the counselors seemed to want to talk with me when I attempted to spark up a side conversation with any of them. The way they avoided my questions and innocent curiosity was similar to my father. I decided to ignore them for now and focus on getting to know my peers.

There were six other boys in my bunkhouse. Guy was a quiet, chubby boy from Israel. Amin was a Palestinian from Jerusalem and had his hair cut like a football player in a style we called *kaboria* at school in Deir. Daniel, an African American raised in Brooklyn, was the first black person I had ever met. He and I connected immediately. When I spoke to him, he would explode in laughter, and when he laughed, I laughed. He seemed to find my every move or gesture hilarious.

"You always look so happy," he kept telling me. That made me cheer up.

"I am a happy person," I lied. "And always have been." In my world, I would not have described myself as happy. I was always overwhelmed by so many emotions, ideas, fears, and aspirations.

There was another Youssef. He was from Egypt and seemed to spend his entire time trying to trade beds with someone else. As soon as he switched beds, he would want to change again. Nabih, a

Jordanian swimmer, was there just to swim, it seemed. Shabi slept on
the bed above mine; he was always singing and smiling at someone.
Tariq was an aspiring actor who found me very funny. He laughed at
my Egyptian accent and my impersonations of Egyptian movie stars
like Adel Imam and Saed Saleh.

The first day started with a gathering together of the whole camp,
or "line-up," as it was called. The first thing we did was listen to
someone from each delegation give a speech. I wished I had been
asked to speak, because I knew I had an important story to tell, but
everyone who spoke was really impressive. There were many journalists and other well-dressed people around. We were called "Seeds"
and all wore the green Seeds of Peace T-shirts we had been given. It
was easy to feel like a football player about to play a really big match.

Flags were raised and the United States' national anthem was
played, followed by the anthems of Palestine, Israel, Egypt, and
Jordan, and finally the Seeds of Peace anthem. We were told, "This
will be your anthem while you are here, and we will sing it every
day." And we did.

The anthem starts with "I am a seed of peace, a seed of peace,
a seed of peace," but at the time I was too excited to keep up with
what they were saying, so I just moved my lips up and down for the
most part. I was aware that the other Palestinians spoke much better
English than I did, and was starting to regret that I had not paid
more attention to my father.

Never in my life had I seen so many girls at once, and they were all
so beautiful. They were all different from one another and yet I liked
them all. However, whenever I looked at them, I would hear my
father's voice: "Stay away from the girls. Stay away from the girls."
So, I would move on and try to focus on other things.

The female Seeds usually giggled when they saw me for the first
time. I never knew why, but they would just laugh and giggle. They
liked to run their hands through my curly hair. If I was eating, they
asked me if I wanted anything more. It always made me a bit shy, the
way they were so kind to me.

The girls at camp impressed me. They were pretty and full of character. At first when I tried to talk with them, I found myself at a loss for words. I felt that I should talk with them like James Bond would. All of that vanished after they got closer to me, especially when they touched me.

Ashley was an American Seed who looked like a princess without a care in the world. The way she hugged me made me feel like I was a hero who had saved Noah's ship, and that made me laugh. Nisreen was Palestinian-Israeli and was part of the Israeli delegation. She was tall and beautiful. When she sat beside me, she offered me one of her headphones. Together we listened to the sweet voice of Fairuz, a Lebanese singer loved throughout the Arabic world.

I wished I were a bit older so I would know how to handle those girls' vibrating energy without feeling so stupid. Thankfully, it was easier interacting with the older women at camp.

We were assigned to groups. I was in Group F, which I liked to think stood for "flower." The other members were Adam and Ashley from America, a Palestinian girl from Jerusalem who beautifully engulfed herself in a *hijab*, a short boy from Jordan who hardly said a word, and two boys from Israel—Gal and Daniel. They were so full of self-confidence that whenever they spoke I felt that I was listening to future prime ministers. Their willingness to express themselves openly put my shyness at ease during dialogue sessions.

The dialogue sessions were my absolute favorite part about the camp. Every day for two hours we met and discussed the big issues that confronted us. Despite the number of times we disagreed with one another, we had a loving relationship between us. It made me love my father with an even bigger smile in my heart.

It was interesting to see Israelis again. Having spent time in an Israeli hospital, I was comfortable interacting with them at camp. In fact, being at a camp for peace was similar in many ways to being at the Sheekum. The Palestinians and the Israelis all had stories to tell, and everyone had different opinions about why there was so much hatred and fear between our peoples.

My father had already taught me a lot about peace, so I could speak with ease on that topic. Sometimes I suspect I sounded like I was giving a sermon, trying to teach them all a lesson. I was becoming a peace advocate or, as some might say, a "seed of peace." I made myself a couple of name tags and stuck them to my T-shirt. One of them had "TNPM" for "the next peace maker."

In my head, I kept hearing the words of my father, "Spread the word of peace." I knew that his commitment to peace was the real reason I was here at camp. Speaking his words was a way I could honor him. Sometimes I was shocked to hear myself defending his message so strongly when others were so vigorously challenging it.

When the Seeds learned that I had been shot, they were surprised to hear me preach forgiveness. They could not understand why I could forgive people who had caused me so much pain. As more people became aware of my story, they were curious to know the details.

The facilitators, one Palestinian and one Israeli, managed our debates with great skill and diplomacy. I cannot remember their names, which may be what they wanted because they hardly ever spoke about themselves.

"Are you the one from Gaza who speaks English?" Ashley asked me as we stood in line at the cafeteria on the first day. It was not long before relationships with my fellow campers became more intense. Although I liked the attention, it was hard when so many of them disagreed with me and so many of them were competitive and committed like me. Everyone held different opinions, but when we were playing sports, all that was forgotten. We became united through our competitiveness. I tried to remind myself that everyone had their story and everyone carried some pain from their past, so of course our realities would be different and we would not immediately agree.

I was even more confused to find that, even though I was in a magical place surrounded by beautiful forests and lakes, my mind was more troubled than ever. I wondered why I had to grow up in a country enmeshed in so much conflict. I wondered about the meaning of the bullet in my back, about the reality of my dreams, and about the wisdom of my father's beliefs. I had never anticipated that one day I would be speaking so forcefully for peace and would not mind how hard it is to do so.

Every morning, we woke up, made our beds, and marched to the morning line-up. Boys sat on the left and girls sat on the right. Every bunkhouse showed up with some form of a walk-in performance. At the dining hall I discovered traditional American pancakes, on which I put just the right amount of syrup. They were like *gatayef* but without the filling.

I was assigned to doing all sorts of activities except for some contact sports because of my injury. Many of the sports were unfamiliar to me. I did what I was told to do, though I often got distracted by not knowing the rules of the game, especially baseball. Baseball and I did not get off to a good start. After waiting endlessly one day for some action, I just dropped my enormous glove onto the grass and complained that my position was boring.

There was so much to do so it did not matter whether I enjoyed a sport or not.

I felt a sense of pride every time Group F successfully completed a given challenge that required each one of us to truly trust one another, if only for a moment or two, in order to survive.

The counselors were all fun and loving. Counselor Eric, always calm and with a smile on his face, was in charge of canoeing, which I enjoyed. I lost myself in whatever game we were playing, despite the nagging pain, and forgot about all the conflicts of the world.

One of my favorite sports turned out to be floor hockey. I had never heard of it before, but now I wanted to play floor hockey all the time. Iman was one of the counselors. I thought she was the most beautiful woman I had ever seen. My heart ached to tell her that, but I was too shy to say anything.

"Are you ready, Yousef?" she would ask with her blue eyes shining at me under her blond hair.

"But I don't know how to play," I stated after some hesitation.

"It's easy," she encouraged me as she handed me a hockey stick. "I'll show you." Then she stood very close to me and helped me position my hands on the stick. A couple of times, she led me through swings of the stick to connect with the puck, which I thought was a very funny word. Slowly, I learned the rules. When I made a good pass, Iman shouted, "Way to go, Yousef."

Once during a floor hockey game, Daniel shouted "YEAAAAH!" so loudly that he made me explode in laughter. That caused me to miss an easy shot. He laughed when I impersonated him, and soon everyone in our bunk was shouting "YEAAAAH!" It drove the counselors insane during sleeping hours.

Things were going along just fine until one day, as I raised my hockey stick to take a free shot at the goal, I collapsed. Suddenly I could not move my right leg. It just shut down and I was in serious pain. All I could remember later was being carried to the infirmary, and the anxiety I felt.

I stayed in the infirmary for a night or two. In spite of the painkillers I was given, my right leg felt like it was getting electric shocks, or as if a piece of fine glass were slowly being inserted into my thigh. I can never exactly describe the brutality of that pain. I focused on the voice inside me that was urging me to use the pain I had endured and turn it into something positive. The voice of my father flooded my very core: "Think of it as an opportunity."

Now, at least, I had the other Seeds ready to listen to what I had to say. *Use everything God has given you. Play with all the cards you have been dealt.* My collapse was a kind of an ace.

Before I could, however, the Palestinian delegation refused to go to any activities in a show of support for me.

What good is that going to do? I wanted to ask them. *It was a sports accident. Football players have them all the time. I was carried off the pitch.*

I feared that I was losing control of my story, that it was having the wrong effect. I loved knowing that people wanted to hear about what had happened to me, but I did not want that to separate us all further, because then I might lose more than I had already lost.

At the next dialogue session after I was released from the infirmary, I described for them how I had been shot. I wanted them to understand why I was uncomfortable hearing about what was happening with my delegation.

I spoke as passionately as I knew how. "The bullet constantly reminds me that I am a human being and in that I find optimism. I have never felt hatred toward anyone and I never had to wrestle with hatred until I was shot.

"For me, believing in peace is not just an act or a few words. I have to forgive the soldier and to forgive the people of Israel. That is how I plan to win back my freedom and homeland." The other Seeds were very quiet, especially the Palestinians.

I told them that I had seen my father confront the soldiers of Israel with his wise tongue and bare hands. I was now making my own calls in life and making my own justifications. I knew that I was serving a higher purpose now: the purpose of my father, the purpose of peace. It would grow along with me as I grew every single day.

After I had finished, nobody had anything to say. I did not know if that was a good thing or a bad thing.

15

Return

Every time I had spoken during the dialogues, I sensed that I was the one with the real power. The soldiers back home were occupying my house and land, yes, but they were not occupying my mind.

Even so, I was worried about going back home. All this anxiety was making it hard to sleep at night. Sometimes Adam would hear that I was awake and come to my bed to check on me, but I would assure him I was OK. I never told anyone about my internal struggle at the prospect of returning home. No one had a clue that I was so upset and so uncertain of my future.

I absolutely had to find a way to study in the US. I had to have more than just three weeks in that country. I learned that I needed to speak to Bobbie, a short older woman with round glasses whom I had seen taking photos of the activities at camp. I learned that she was the cofounder of Seeds of Peace. I felt intimidated at first, as I thought she might not take me seriously.

I finally managed to break through my fear and go talk to her.

"Hello, may I speak to you for a few seconds?" I asked as I tried to stop shaking. I told her about my dream of staying in the US, and attending a boarding school to prepare for university.

"A lot of Seeds want to do that," she said kindly. *Of course they do, but I am the one with the bullet,* I heard myself silently saying. I felt like I had the bigger story.

"Seeds of Peace is not in a position to pay for campers to attend schools in the US," Bobbie added quite emphatically. "But I will keep it in mind. You never know what might develop."

The next week, she had me meet with some visitors to the camp, some of whom were university presidents and secondary school headmasters. This encounter only made me believe all the more in my dream of studying in the US. Although she had said she could not help, she was helping.

All too soon, it was time to pack and head back to Gaza. Though we had mostly exchanged laughs, Daniel and I suddenly one night found ourselves in the midst of a long, emotional conversation. We spoke as if we had known each other from another time. Though he is Christian and I am Muslim, we uttered almost exactly the same words. We shared our hopes and dreams for the future and experienced that special warmth that comes with honesty.

I secretly envied him because I was about to head right back to Gaza while he was going to Brooklyn. I felt a sense of gratitude for having met Daniel because the soldiers back in my house had generated such a negative impression of the people of Brooklyn. By knowing Daniel, I was reminded that goodness exists anywhere and everywhere. Sometimes it exists right alongside evil without us even knowing.

It amazed me to see how the Seeds, who at times had disagreed so vigorously with each other, were now all crying as we said goodbye. Israelis, Palestinians, Egyptians, Jordanians, and Americans—all of us were in tears. Even the counselors were crying. I, too, was crying. I did not want to go back to my "jail house." I wanted to scream those words at the top of my lungs.

When it was time to leave, I exchanged as many hugs as I could with Seeds, counselors, and especially with Kelsey, an American Seed. She had treated me with a kindness that I will always remember. I also gave a big hug to a counselor named Aaron who had given me his raincoat at a baseball game on one of the last days of camp when it was pouring.

As we got closer to the airport and to the final goodbyes, I felt increasingly alone. My father and the soldiers had each in their own very different ways helped me reach this beautiful place. Both had given me a mission and had started me on a spiritual journey, but I wondered how I would find a way to continue that journey. Maybe everything that had happened at camp did not really matter if, in the end, I did nothing more than go home and surrender myself once again to the occupying Israeli army. There must be another way out.

* * *

We flew off to Cairo. I shut out everything, feeling worse by the hour. We finally arrived at Cairo's airport and fell into a big mess. The immigration officer looked at me coldly. "Where is your visa?" He said the same to another one of the Seeds, whose name was Mohammad. We waved the teachers over to where we stood in front of the immigration desk. The officer did not treat them any better.

"They are over sixteen," he said flatly. "They have Palestinian ID cards. They are treated as adults. Adults need visas to enter Egypt. They cannot enter Egypt without visas." Visas? We had no visas.

While the other students in the delegation, who were minors, continued along with the female teacher without us, we were forced to stay behind with the male teacher. This was the first time I had been pinched by the lovely Egypt. The land of Joseph and Moses had asked me, a son of the Holy Land, for a visa. The land of Arab unity and the land of Arab glory had asked me for a visa, which I did not have.

"Follow me," a policeman said. He herded us into the dirtiest room I had ever been in. It looked like some kind of large bathroom. There was one chair, but it was broken, so there was no way to sit except on the dusty floor. Several other travelers were there already, looking godforsaken. The two officers posted just inside the door kept talking about what they were going to do that night if they could find women to do it with. Their conversation was slightly interesting to listen to for a while, but it seemed strange for them to be so open about it. They spoke as if we were not even there.

I felt bad for the teacher, because they kept giving him a hard time. I tried to be helpful.

"Look, sir, if you just pay them something, they will let us go."

He blew me off, so I went to sit in a corner and watch the cockroaches run in and out of the wall cracks.

I did not mind sleeping on the floor; I had to do that at home. Only this time I found myself observing strangers I had never met before and was unlikely to see again. The only thing we had in common was that we were being held in that prison-like room waiting to be deported back to Gaza.

The next morning, we were finally allowed to cross the border. People had been making phone calls to embassies and various authorities all night. My father met us. The way he hugged me and threw jokes at me showed how happy he was to have me back safely. Then he asked the teacher, "How was Yousef? How did he do?"

I do not think the teacher even gave him an answer. He was too busy taking his stuff out of my luggage to answer. His bag had been overweight, so he had saved money by putting some of his things in my suitcase. Immediately, though, I started to feel the same old pressure I had always felt when I was with my father.

As I arrived home, Ben Wedeman and his CNN crew were there. They filmed me as I entered the house. I even did a live split-screen interview on CNN along with another Seed, an Israeli, who was already at home in Tel Aviv. That moment was very special to me because although I still had some trouble speaking English, I felt that a window had opened up for me to communicate with the whole world.

My mother, brothers, and sisters were all happy to see me again—especially my mother and my youngest sister Zana. My mother had prepared a giant meal for me. On the table were stuffed grape leaves, salads, stuffed and roasted young pigeons, and that was even before getting to the main course: a whole platter of *maqlouba*.

Everyone else in the house was glad to have me home. I was the only one not happy to be there. All I wanted to do was turn around and head the other way. I was determined to find a way out.

Within about two weeks, I received an e-mail from Bobbie. It turned out she had not forgotten how much I wanted to go to school in the US.

"I can't help you come here, Yousef, but there is a Quaker school in Ramallah in the West Bank," she wrote. "I am trying to get you admitted there. Does that interest you?"

"Yes! Yes! Yes!" I quickly wrote back.

"Good," she replied. "I am just worried to bits that if you stay in Gaza you'll end up getting killed."

Bobbie, who was not a Quaker but had a profound Quaker connection, and Eric, the canoeing counselor at the camp whom I had really liked, had gone to that school. They both had spoken to the headmaster about me. Bobbie offered to pay my tuition fees. Going there, though, depended on whether I could get a permit to leave Gaza. Again, I pleaded with my father.

"If you want me to get As, you should let me go to school in the West Bank," was how I opened negotiations, even though the real reason for going was just to get away from the soldiers and all the craziness they had brought to our lives. He said nothing.

"I will be closer to Tel Hashomer for checkup appointments," I told him a couple of days later. More silence. No argument seemed to work.

I was back to being ordered into the living room every night by the soldiers and being told to study by my father. I would go to a corner in the house and just sit there, staring at a book. I looked like I was reading, but in reality, I was just thinking about Maine and Boston and all the great things I had felt while I was there.

* * *

The soldiers were suddenly acting as if it were the last week of school when no one cares about the rules anymore. For the first time I had a real conversation with one of them, as he sat in the hallway guarding us.

I told him about Seeds of Peace and even tried to interest him in going there, although he was too old. I fantasized that maybe I could transform him from a soldier in my house to a peacemaker at Seeds of Peace. After my six months in Tel Aviv and my three weeks in Maine, I did not fear this man because he was a Jew or even because he was a soldier. He was just a man doing a job he did not like. I could see that, and he knew that I could see it.

"If nothing else," I pointed out, "you will get away from all this for a while." I made a broad gesture with my hand. The way he said "Yeah," I knew he wanted to be out of there every bit as much as I did.

"Take this. I want you to have it." I handed him one of my Seeds of Peace T-shirts.

"Thanks," he said as he looked at the design on the front. "Maybe I'll go some day." I did not know if he meant that, or whether he ever thought any more about it, but at least he had made the effort to talk with me.

On 7 August, the very day that the TV announced that Israel had ratified a plan to withdraw unilaterally from Gaza, my father told me he had received a permit for me to cross at Erez. I did not pay much attention to the news reports. I took it as only another idle promise. Instead, I got revved up about moving to the West Bank.

The announcement of the withdrawal, however, was true. The Israeli settlers at K'far Darom, our neighbors, were told they had to leave in the next week. Large numbers of soldiers started arriving to help them pack their belongings.

I was told to go pack my stuff, too. It had not been easy for my father to get me the permit to cross the border. You never knew when the border might suddenly be closed for a few days, or a month, or a year. Once somebody had a permit, the important thing was to leave as fast as possible and not get stuck.

I felt weird about going away just as the soldiers were also leaving. I had prayed for years for them not to be there, but it had never occurred to me that we might end up leaving at the same time. I had dreamed for so long of going to the beach again, and now, just as it was becoming possible, I was moving away. I was determined, however, to put any feelings of homesickness to sleep.

Many of the settlers resisted the soldiers and tried to stay. The soldiers treated them very gently, so differently from the way they had treated my family who had never fought them. We watched on TV as some of the settlers barricaded themselves in their synagogue near our house and attacked the soldiers with eggs and cans of paint. The soldiers sprayed them with water cannons. No live rounds. No rubber bullets. It was all over in a couple of hours. Then, as the settlers left, the soldiers bulldozed their houses so no Palestinians could live in them.

The CNN crew had come to cover the eviction of the settlers. They had been such frequent visitors at our house that when they heard where I was going, they offered to take me with them back to Erez and then on to Ramallah. Their media credentials meant we would not have to wait at the checkpoints. I hated waiting in line at checkpoints, especially after everything I had just experienced in America.

There was one last comic drama that had to be played out before we could go. I can still picture my father standing at the living room door, arguing with the guard who would not let me out of the house. The CNN car was parked in the driveway ready to take me, but we had been ordered to stay inside for the duration of the Israeli evacuation. The soldiers did not want any ugly incidents.

"My son wants to go to school in Ramallah," my father said, very quietly, but very passionately. "He went to Seeds of Peace, and he wants to work for peace. You must let him go," he said.

"Sorry, but that is not possible." The soldiers were adamant, but they had used the word "sorry" for the first time since they had occupied our house.

"I must speak to your commanding officer," my father went on, getting no results.

Suddenly, everything changed. The soldier to whom my father was speaking got a message on his walkie-talkie. He turned to my father and told him, "Get him out right now."

My father picked up my bag and almost threw it in my arms.

"Go, Yousef. Go!" he exhorted me.

I wanted to hug him and tell him, "Yaba, you're a hero. You've won. You're finally getting what you've prayed for. You're going to have your home back, your land back. You'll be able to live in peace again."

I wanted to tell him how much I loved him, but I could not bring myself to say any of it. I felt too shy in front of the soldiers, my siblings, and the CNN crew. I might have been able to tell him if we had been alone. I quickly hugged my mother, my sisters, my brothers, and finally my father.

I could feel his anxieties about the soldiers leaving. He did not celebrate the event, instead he only wondered about the future. He hoped it would lead to real peace in our land. Even though the Israeli army was packing and going away, they still controlled the borders. Air, land, and sea.

I had felt so squeezed by my father's way of life for so long, and I had felt so squashed at the way he made us remain in the house with the soldiers. Now, though, I was so deeply proud of him. I knew he was a hero, a real hero. He had remained true to his beliefs, and by his example he had taught us, his children, to do the same. I was able to give thanks in a new way for how he had raised me and for his way of life. He was a true soldier of peace. He had not allowed his hopes to die and, in the end, he prevailed.

"We will never become refugees." How many times did we hear him say that? I was thankful that I had not become a refugee and that he had been able to save his home and his land—the home and land I was now leaving.

"Go, Yousef!"

16

Ramallah

The moment I arrived in Ramallah, I knew that things were not going to be easy. My father had arranged for me to live under the supervision of my aunt, who would also serve as my official guardian. I had always liked her. She was a short woman with a big personality like her mother, Sitie. She loved to laugh. It was clear, though, that she did not share my belief in peace or interest in reconciling with Israel. She had her reasons.

I tried hopelessly to explain my views to her. The more I explained, the less pleased she was to have me with her. She constantly belittled me. She saw no future for me at the Friends School and certainly not in America.

She had been living in Ramallah since 1980. Her husband had a business there. As it became harder and harder to get permission from the Israeli army to cross between Gaza and the West Bank, they decided to settle down in Ramallah. I am sure she must have thought that things would get back to normal in a few years, but they only got worse, and Ramallah had become her home.

I quickly discovered at the Friends School that whenever I started talking about coexistence everyone kept telling me I had been brainwashed and was crazy. Even some of the teachers. Many of them were Palestinians who had no interest in coexisting with Israel. I did not blame them for that. I knew what we had all been through at the hands of Israel. I hoped, in return, they would not judge me. Many carried foreign passports; some of them even had US passports and could travel wherever they wished. They, of all people, should have seen the need for a peaceful solution. Yet, I felt like I was the only one at the school who understood the importance of peace. I, with only a Palestinian Authority passport that for five years had barely allowed me to go to my own bathroom, seemed more aware than all the rest of them. It blew my mind how so many Palestinians walked around with hatred toward one another because of a mess started by the British, the Egyptians, the Israelis, and a whole lot of other people who were not Palestinians.

As I came to understand how hard it was going to be for me in Ramallah, I returned to my old fantasies. I dreamed that an angel would suddenly appear who would whisk me away to America. Surely, I would find more support for my ideas there. I even began to imagine being accepted to a boarding school in the States. It did not matter which boarding school, only that it was in the US. I pictured myself walking into a classroom and feeling welcomed by everyone.

My aunt had rented a room for me with a group of young Palestinian men who worked near her boutique. It was definitely not what I wanted and certainly not something I enjoyed, but I kept assuring my father I was happy. I did not want to say anything that might cause him to come and take me home. Because the other guys in the house had to go to work early, I got up early with them. They were less concerned with my activism than they were with making enough money to support their families. They worked all day and still complained that they could barely make it, and I did not doubt them. They ate pasta for breakfast, lunch, and dinner. I could see that they were the ones who were really getting screwed by the system.

I would wake at 6:00 AM and walk the three miles to school. It was all hills and mountains, and I loved every step and every new view I discovered in my Holy Land. I walked both ways because it was my first time in Ramallah and I wanted to get to know it. Also, I wanted to save the taxi money and keep it for things I needed.

The Friends School was established by Quakers, or "Friends" as they are formally known, and some of its teachers were from the United States. It is the oldest school in Palestine and has been around since the time Palestine was under Ottoman rule.

Gradually, I became friends with several of the Americans at the school and their families. The university counselor, John, was from Kentucky, and he sounded like it, though his parents were Palestinians. Whenever I knocked on his door, he would answer with a smile and that harmonious Southern accent that I had never heard before, "Yousef, what's going on? Com'own in."

One day I got up the nerve to go in and sit down and tell him my story. I pleaded with him, "John, I want to go to school in America. Help me go to a school in America."

I could tell he was interested, but it was also clear he was unsure how to help.

"It's hard, Yousef. It's very expensive to go to school there. You gotta understand."

Every day I would knock on John's door with the same plea, "Help me get to America!"

I bothered John so much about going to the US that I drove him mad. It was almost as though I were stalking him. It got to the point where, at the end of the day, he would sneak off and hide until I had gone home. I could not stop thinking that he should be able to help me. I was also tormented by the thought that if I was not accepted to a school in the US by the end of the school year, I would end up having to go back to Gaza. If that happened, I would have accomplished nothing.

Once, when I was in John's office bugging him about going to America, he handed me a list of US prep schools from a magazine and suggested I contact them and tell them my story. Maybe one of them would take interest. I was ecstatic! I started calling and writing American schools and attempted to persuade them to admit me. I was sure that if I kept knocking on doors, one of them would open.

* * *

Although I was slowly making friends with some students and beginning to get along with some of my teachers, I had no one with whom

I could simply be myself. I found myself starting to tune out everything that was happening around me. If no one wanted to sit down and get to know me, I would simply ignore everyone. It was disillusioning to find that no one seemed to share my passion for peace, not even in a Quaker school established to foster peace. Whenever I tried to talk about the human side of the Israeli people, I would be accused of betraying my own people. During one class a teacher even shouted at me, "Yousef, you will become a puppet for the Americans!" I took his words as if they were compliments.

The more I tried to smile and explain myself, the more everyone made fun of me. My emotions were squeezed, but surrendering was not an option. All I needed was support from the significant people in my life—my father, Bobbie, and John—to help me get what I wanted. The rest did not matter to me, and so I ignored any discomfort my peers caused me.

It had been a while since I had seen Omar and the boys. If I had told them how I was no longer so popular at school, they would not have believed me. I missed all the stunts we used to pull together. No matter how many times we got busted and apologized, we would find ourselves involved in some other situation in no time. At my new school, though, I was no longer in the mood for pranks. I needed to focus on my future.

I was finding it difficult to sleep at night. The guys where I was staying went to bed very late, and I had to lock myself in my room to get any quiet. The pain in my back was constant and made it hard to sleep.

My father was ringing my mobile every day to check on me. Every time I heard his voice, I wanted to tell him how bad things were, but decided to reveal only what I thought he should hear. I dared not tell him that I prayed every day to attend a school in America. However, I did tell him that my back was often unbearably painful.

It felt as though electricity were flowing through my back and down my legs. When he heard that, he quickly arranged for me to have regular physiotherapy at the Sheekum in Tel Aviv. Those trips soon became some of the best parts of my week. I got to see where I had stayed and to check out the shawarma joint, although my friend Nicky was no longer there. I wondered about Mohammad and what had happened to him after he moved back to his home in the Nuseirat refugee camp. I could visualise Miko in a speeding car with the speakers pumping.

Miko and Mohammad are going to be in wheelchairs for the rest of their lives. Perhaps that is why we never bothered to keep in touch. Perhaps we knew the truth about everyone and everything. I was walking and could engage again in the normal life around me. I could adjust to my situation while they had to adjust to being disabled. I felt I would be embarrassed if I ever saw them again.

On these trips, Rami was the taxi driver who took me to the Israeli military base, where I had to go to get my permit to cross into Israel. Then he would take me to the crossing point at Kalandia.

"Yousef, ya Yousef, ya Yousef," he would sing when he saw me. He was a young Palestinian man who seemed fully content with his life, totally satisfied to have two Nokia mobile phones attached to his belt and his own taxi with which to make a living. I knew that lifestyle would never be enough for me. I could think of nothing else but being in the US. I shared my dream with everyone, even strangers, hoping for some words of support. Any time I shared my dreams with Rami, he never let me down.

"I want to go to America, Rami."

He would laugh. "You look smart. I'm sure you'll get there." We treated each other like an older brother and a younger brother.

After he let me out at the checkpoint, it could take hours and hours to cross. Long lines of Palestinians were forever waiting. That was not a pleasant experience. Everywhere you looked, you saw fences. Everything looked like we were in one big cage. In fact, it was exactly that. It was as if *The Pianist* movie were being replayed, but in real life. In my life.

Once I reached the other side of the crossing, I had to catch a bus to central Tel Aviv near the hospital. I would observe the Israeli civilians. Young women in tight jeans. Boys and girls laughing and shouting about something. I kept waiting for someone to give me a dirty look because I was an Arab, but no one ever did. My impression was that whoever checked me out decided that I was Jewish.

Sometimes I watched the military trainees as they boarded the buses and began playing around with their large M-16s, most of which I could tell were not loaded, since I had become a secret expert on guns.

When I finally arrived at Tel Hashomer, I always looked first for Seema, but I never found her, though I was told that she was still working there. Dr. Brezner would be waiting to greet me with his usual smile as he gave me a checkup. Apart from him, I never saw any of the others who had cared for me.

On the way back to Ramallah, I often stopped to eat at Pizza Hut, which was not available in the West Bank or Gaza. Sometimes, though, I got a ride from Eric, who worked for the Seeds of Peace office in Jerusalem. His car had an Israeli license plate, so if I rode with him I would not have to go through the hassle of crossing checkpoints. I would not even have to show my permit or even my ID. This made me feel embarrassed: every time we drove by, I would look at the women and men just waiting under the sun for the soldiers to let them through. I never said anything to Eric, but I was plagued by my guilt. The Israeli cars had full access to all destinations in Israel and the West Bank, just as they did when they were in Gaza. They even had their own roads and highways.

Despite the checkpoint hassles, I liked going to Tel Aviv, because I could enjoy a few hours away from the Friends School. By then, my marks were falling and I would try to comfort myself by thinking, "You are failing because you are here. You are a good student. You will do well somewhere else. You must find a way to go where you will succeed."

In class I started to taunt the teachers who were bothering me. I simply did not care anymore, and I enjoyed my apathy. They had awakened my appetite for pranks. If my teachers had had sons who treated them as I did, they would never have become teachers. Even though my Palestinian teachers and I were from the same country, it felt as though we were strangers. I might as well have come from the moon. The war had divided us for so long that we no longer seemed to share the same nationality. My classmates considered themselves among the elite of Palestinian society. To them, it was obviously much better to be from the West Bank than from Gaza.

I disliked how there was so much animosity among the Palestinians. Everyone was marked by their background—one person was a refugee and another a native, one was poor and another was rich, one was from a village and another from a city—not to mention what kind of job they held or what family they came from. This was how everyone around me seemed to judge other people. My people cannot afford to abhor one another, but loathing is what I tasted in Ramallah, wrapped in a delicious shawarma from the vendors in Al-Manara Square. We can never defeat the occupation this way, and that was how I usually justified everything I was doing during my life at the Friends School.

17

Desperation

Students at the Friends School had the option of making a presentation once during the year at the weekly school chapels. The outspoken ones took advantage of this. I did, too. When it was my turn, I spoke about my newly formed philosophy of peace.

I was nervous, and frightened that I would show it. I had to be strong to wake these students and these teachers from their deep comas, induced by their preoccupation with their social status. I gripped the lectern in front of me tightly and started. Of course, my voice squeaked when I began to speak, but that forced me to take total control of myself and what I was saying. I could not allow myself to think about anything else.

"I am historical Palestine, modern Palestine, and future Palestine," I started. "I own land in the Holy Land and I have to travel to another city to meet someone who does not know my family's name.

"The Israelis have occupied our lands and destroyed our country, but they have not destroyed our humanity. Our purpose as people of the Holy Land is to restore the values of peace and to foster forgiveness, because it is only through coexistence that this land can flourish again.

"It is the peace that we must now win, not another war in which we are betrayed again by all and taken for granted again by the entire world. We will not defeat them in the same way they have defeated us, because we are people of peace, not war, and there is pride in that. We can protect our land and human dignity by showing them and the world that no one else ever can or will care about the land as much as the sons and daughters of Palestine. Our land is holy and our land is blessed.

"My family stood up in the face of occupation by not leaving our land, by not allowing fear to control our destiny. So, today, I stand before you thankful for still being able to point to my home in Gaza on the world map. My father defeated their weapons with words of love and peace. I intend on forgiving them because that is my destiny in this world and that is the purpose I shall serve.

"A soldier shot me for no reason other than my being a happy Palestinian boy. I choose to forgive him, not only because his people were the ones who saved my life but because forgiveness enables me to think about the future of my country and the future of my people. I am a person of forgiveness because I am a Palestinian and I care deeply about the goodwill of this land.

"This hatred of ours toward them is nothing but a delusion, for it only leads to our destruction and misery. It enables them to describe us in the most awful of ways to the rest of the world.

"We are lucky to be attending a private school here in Ramallah, but you and I know that many others have to walk to school from their refugee camps. While we are living well here, we have access to Israel. Your fathers would not be able to afford to send you here if Israel did not exist. Many of you have American or British passports. You all live your lives as if you were in fact from America or England. And yet, you dismiss the idea of making peace with Israel, even though it will help those who have been pushed out from the life of privilege.

"It will never be a shame to want and wish for peace with the people of Israel—they are the sons of Abraham just like we are. They may choose to hate and persecute us, but we will never go away. We shall always be here to remind them of the right path. We shall share our stories of pain and suffering with their women, their children, their youth, their fathers, and their mothers. We shall keep on telling our stories to them and the world, for we are holy people and this world will never be at peace if its holy people are being persecuted without just cause.

"They have taken our land, but we must not let them take our souls and hearts. We are people of peace and compassion and therefore we are the people who have the large and holy duty to restore peace and justice. We shall always have our hands extended for peace and we shall pray to God for patience. We shall stand up for our rights through the words of peace, for they can never win."

There was a cold and graceless silence as I finished and went back to my seat.

A few days later, the physics teacher stopped me and said, "You are not from around here. You will never be a leader here." I wanted to tell him, "I have no desire to be a leader of my people. Not now. But if I can find a way to go to school in America, I might learn how to achieve peace for Palestine. Then we'll talk."

As the weeks passed, my marks dropped to Ds and Fs. The school put me in the classes taught in Arabic instead of English.

"You're wasting your time sending him to that school," my aunt told my father. "You should take him home."

"She's just saying that to get rid of me," I told my father. The truth was that my relationship with my aunt had deteriorated drastically. When she complained to my father about me, he asked her if I had stolen something, smoked, or done something bad.

"I could live with that," she told him. "No, the problem is that all he talks about is Seeds of Peace."

"You're not a real Palestinian," she spouted every time I shared my views with her. What she said stung, but in the end it did not matter to me. I just reminded myself, *I'm going to America.*

One Friday the headmaster sent for me to come to his office.

"Your aunt just phoned. She says that she is no longer willing to be your guardian. You know that the school has a policy that a student has to live at home or have a guardian."

"Yes, sir," I said, not knowing where this would lead.

"I am not sure how we are going to be able to keep you. Since your home is so far away, this creates a real problem."

Luckily, I had a good relationship with our headmaster, who was

Christian and knew my story. He had always been very kind to me and sometimes I had even cried in his office. One day, after I had told him about a student who had been unkind to me, he got so upset he ran up the stairs to the classroom and shouted at my classmates, "Shame on you, leave him alone." I will never forget how he stood up for me. The class would act sympathetic for a few days, but then everything went back to normal. I felt like a coward for all my tears.

My father and Omar better not learn about this, I said to myself.

* * *

Every day I e-mailed prep schools, but none responded to me. I was getting very discouraged. I had no guardian and no place to live. My aunt's husband sold Range Rovers, so I would go to his dealership and sleep beneath the new jeeps that were parked outside. When police patrols came at night to do a security check, I would snuggle up under a car and hold my breath until their lights passed. For a couple of days, I enjoyed living like that. I felt like I was on an adventure and finally in charge of my life.

It got tiring pretty quickly. So, I took my things to John's flat and insisted that he host me for a couple of days until I could work things out.

John told Tim, one of the teachers, about my situation. Tim invited me to move in with him and his wife Lisa and their young son Paul for the rest of the school year. Tim spoke very good Arabic. Paul was a little blond boy who was always playing a Spider-Man video game. He was extremely polite.

They lived in a beautiful flat just across the street from school and had a garden that was maintained daily by the school. It was a great place to stay. I learned later that many of the faculty had advised Tim against taking me in, but thankfully he had ignored them.

The truth is, Tim saved me. While living with them, I learned a lot about what it was like to be an American, as well as what it meant to be Christian. Tim was originally from Nebraska. He and

Lisa had married when they were teenagers and had been together ever since. They often prayed together. To understand them better, I went to church with them sometimes. I am very proud and happy to be a Muslim and will remain Muslim for the rest of my life. The message of Prophet Mohammed, peace be upon him, is one that I fully embrace. However, I was curious about Christians, whose Prophet Issa [Jesus] is an important figure in the Holy Quran. In fact, I must believe in Jesus and Moses and all of God's prophets if I am to be a true Muslim.

Tim and Lisa were Greek Orthodox, and they told me a lot about the different branches of Christianity that I did not know. I was interested in learning more, though my classmates criticized me for doing so. By then, my relationship with my peers was so bad that I hurried out of sight when school ended. Everyone at school knew I had a bullet in my back and that I had gone to Seeds of Peace. Some students bothered me about Seeds of Peace. When I complained to the teachers, they just ignored me. Still, every day I put on my Seeds of Peace T-shirt under the shirt I had to wear for school. It became like my own personal uniform.

My legs were not working well and I would sometimes fall. Sometimes my right leg would become totally immobile and I would find it almost impossible to walk. The first time that happened, everyone got very worried. When it began to happen more frequently they seemed to stop noticing.

I kept taking more painkillers as if they were M&M chocolates. One evening, I swallowed an entire bottle of painkillers and ended up at the hospital in Ramallah having my stomach pumped. When I regained consciousness, John was standing over me and whispering, "What the hell, Yousef? The doctor sayed you trad to kell yourself!"

I knew that the doctor was probably correct, but I just answered, "What are you talking about? I was just trying to stop the pain."

Tim was also upset, and I was very embarrassed. He was shouting at me as though he were my father, "I go to Jerusalem for a weekend, so you can commit suicide?"

My father had to come, but he did not manage to arrive until after I had been released from the hospital. I went in a cab to meet him at the border. He was so angry that he did not even speak to me until he was with John and Tim. I kept reminding him that I had had unlimited access to morphine when I was in the hospital in Tel Aviv, so I presumed that the solution to pain was to take more pills.

Without even listening to me, he lashed out, "You were probably trying to be Valentino over some girl!" Well, that may have been true back when I was writing love poems in grade seven, but I had given up girls when it became obvious that these West Bank elite girls were not interested in me. Besides, I kept trying to assure him that I would never kill myself over a woman. *At least not yet*, I whispered to myself for some reason.

"I was in pain, Father. I swear. Don't be worried, Father. I'm not like that. Besides you have four other sons and three daughters and they are amazing, so don't worry so much about me!" He was too upset to even respond. At the end of the day, having decided I was fine, he went home.

* * *

I never shied away from bothering John about schools in America. "Yousef, you need to focus on your homework," he would say without much sympathy. I treated John like an older brother and would bug him to do stuff with me. I would go to his office and say, "John, what are you going to do this weekend? Do you want to do something together?" We would hang out. He never stopped reminding me, though, that my English was bad and I should be studying. This really upset me. If even John thought I spoke bad English, maybe I would not get to go to America.

I told him that I had already taken the TOEFL exam, but I presumed my score had been low, because it had failed to impress my aunt. I remember going into her clothing shop all excited to tell her about my TOEFL scores.

"Everyone knows English now. It's no big deal, Yousef," was all she could say.

I was upset and tossed the TOEFL report in the bin and went out to get a shawarma. I did not know what was high or low. I had just answered as many questions on the exam as I could.

After I managed to get another copy of my TOEFL score, John was shocked when he had to tell me that I had done well. My score was higher than average.

My birthday was coming up the following week. I wanted to celebrate. I invited all the students in my class to a party. We were not great friends, but a birthday demands a party, and a party needs people. No matter how badly I was treated, I still wanted to make friends there. They were my people after all, so why not?

After school, I ran to the bakery and prepared a large table of snacks, but only a couple of people came: Ghasan, Sa'ad, Sameh, Basil, Subhi, and a few others who could not stay long. Ghasan had gone to Seeds of Peace camp, so we had some things in common and got along all right. As for the rest, I told myself that it did not matter.

The truth was, though, that I now felt as though everyone was upset with me: the teachers, the students, my father, and Tim and John. I remembered something a soldier had once said to me when I wanted to be allowed to go onto the veranda to watch a football match: "Yousef, look at your siblings. They don't bother us like you do. Why don't you do your homework like Khalil wants you to?" He had said it as though my future and well-being were his top concern.

I was constantly reminding myself to be positive and to keep making an effort. I had to do whatever it took to cheer myself up. I knew I had to focus on the future; I was my own best friend. If I could do that, it would keep the spark glowing inside of me.

I would tell myself, *Yousef, it's going to be good. It's going to be good, don't give up.* I knew I had to get somewhere where I could feel some peace, some hope for the future. I fantasized about Boston or Maine, but I was willing to settle for anywhere in America. I just had to figure out how to get there.

18

"Forward, Yousef!"

The dream was the same, night after night: there is an e-mail waiting for me. I try to act cool as I open it, even though I am sure I know what it contains. I feel like I am embarking on an important journey, and I do not want to make any mistakes. The e-mail flashes on to my screen. Yes! Just as I expected. It is an acceptance from an American boarding school. I start reading the letter out loud, but as I get to the name of the school, the letters start to become fuzzy. Then they start moving around and the screen fades. I breathe deeply and try not to get frustrated. I close my eyes knowing that when I open them again, all the letters will be clear. Then I wake up.

I was constantly being kicked out of class. Not that I minded very much, because I did not want to be in class anyway. Everything was far too hard for me to understand and it was torture to sit there and not have a clue what was being done and said. My classmates had been studying in that school most of their lives and I was the new boy in town, hopelessly trying to fit in. I would skip school and go to a coffee shop in Al-Manara Square to play FIFA video games. Whenever I played a match, I would think to myself, "If I win this time, it will mean good luck. I will go to America."

One day it really happened. When I logged onto the internet that evening, the first thing I saw was an e-mail from Wasatch Academy.

"Dear Yousef," it said. "Thank you for your e-mail and interest in Wasatch Academy, the only university prep boarding school in Utah. We think you would be a great addition to our community and would like to invite you to apply. Best wishes."

I shouted like a madman, then ran out into the garden and raced through the trees and flowers, yelling loudly, "YEAAAAHHHH!" I was so happy and so excited and instantly let go of all my grudges. To me, it meant that the matter was as good as settled.

"They want me to come!" I assured myself. Why else would they have written back? I saw John coming out of his office and raced toward him, screaming.

"Yousef, it's just an e-mail urging you to apply," John said when he saw it.

"What? What do you mean?" I shot back as I followed him into his office. "Didn't you read it?"

I was convinced the school had said they wanted me. John shook his head and pressed his lips tightly together.

"They say that to everyone who writes to them," he explained, with the tone of exasperation that I increasingly heard him use toward me. He pushed a desk drawer closed and shifted his focus to his computer screen. The bell for the first class rang and I had to head off to it. As I left his office, he looked up at me and said, "Fill out the application, if you want. But don't get your hopes too high."

I cut classes that morning and filled out all the forms, gathered all the documents they requested, and by noon sent everything off via DHL using the money I normally would have spent on FIFA-playing and shawarma-eating. For the next few days, I spent every minute I could at the coffee shop, using the rest of my money to buy calling cards and internet hours to be sure Wasatch had received what I had sent. I counted the days and the hours, waiting to hear back from them.

It was several weeks before I received a reply. Then one day, in the post, I received a large FedEx envelope from Wasatch. In it was an acceptance letter. They even sent along a list of books they hoped I would read before my arrival. I felt paralyzed again—not because of a bullet but from utter happiness.

I ran to show the letter to John as he was just coming into his office.

"That's great," he said as he focused his attention on a pile of papers on his desk.

"Aren't you excited?" I practically shouted. "Get excited, John."

"OK, I am excited," he said, with no trace of excitement. "I'm also very worried that in the end you are going to be terribly disappointed." He sat down and looked up at me. "How in the world do you plan to get to Utah?" he asked. "Who is going to pay your tuition?"

I had no idea. It never occurred to me to note that there had been no mention of tuition costs or a scholarship in the invitation to apply. I just knew it had to happen. I was going to go to America. My dream was too clear to not be real. The next step was to go back to Gaza to talk things over with my father. The school year was ending. The timing felt right.

* * *

As I left Ramallah, I was surprised at how sad I was to leave. Hamas had won the elections a few months before and people were very engaged in the political process. It made me smile when I ran into a demonstration being conducted in a civil and democratic manner, or saw one of them on TV. It did not matter to me whether it was Fatah or Hamas. They filled me with optimism and made me realize that I was even going to miss those people who had given me a hard time. After all, they were my people. I felt proud to be a Palestinian.

Doug, one of my teachers, drove me to the border crossing at Erez after the last day of school. As I passed through the border, I could see my father and Uncle Mujeeb waiting for me at the end of the long tunnel.

"*Hoy hoy ya*, Yousef," Mujeeb teased when he saw me. "You've got a blazer on. Big man!" I was still wearing the Ramallah School uniform. "Well, get ready to take it off 'cause we're going home." I ignored him and hugged my father, but his words stung inside. He knew that going home was not where I wanted to be.

"I'm only here for the summer," I declared to myself.

When I got back home, my mother was not there. While I was

in Ramallah, she had learned that she had weak bones and needed to go to Jordan for treatment that would last several months. She had a brother in Amman and was staying with him while she saw her doctors. When I realized that if I went to America, I would be leaving before she got home, I was overwhelmed with sadness. Not only was I desperate for one of her amazing hugs after my nine months away, I needed her because she would have pushed my father to allow me to go. She supported me as long as I promised to come home after my education was complete.

My younger siblings were excited to see me when I came in the door, but after about five minutes, they went back to whatever they had been doing. The house felt the same except for one huge difference: there were no Israeli soldiers there. They were not even in my city. They still controlled who and what got in and out of the Strip, but I could go upstairs any time I wanted. I could go outside and walk around the house any time I wanted. I could come home late.

Yet, where the greenhouses had once stood, there were just tangles of metal that had once been their frames. They lay twisted among weeds that had grown in their place. The soil that my family had carefully built up for generations by composting and careful farming had been gouged open by the bulldozers, exposing the hard earth deep beneath. The land that my father had made bloom looked raw and ugly.

The house was heavily damaged. Now that I could safely walk around the house outside, I could see how damaged it was. In many places, bullet holes had scarred the walls. My father had left his bedroom exactly as it had looked on the night he had been wounded there. It was like a shooting gallery or a crime scene without the yellow tape. On his desk were some of the hundreds of bullet casings and pieces of shrapnel that my brothers and sisters had collected, which had been fired around us and, sometimes, at us. Hanging above them was the peace verse he had brought home from his Sons of Abraham meetings five years before.

Peace begins at Home
Peace begins in Me
Peace begins in You
Peace begins in Her
Peace begins in Him
Peace begins in Them

It still leaned against my mother's mirror. It had made it all the way through the Intifada untouched, though the bedroom had repeatedly been pounded by bullets and small-range Israeli missiles.

Despite the destruction, it all looked glorious and beautiful to me: there was no more shooting outside our windows and there were no tanks parked by the kitchen. No more towers. No sentries outside our door. It was like the soldiers had somehow, mysteriously, disappeared.

As my father and I sat on the veranda drinking tea, he told me about the last day the soldiers were there.

"The night before they left," he said, "they took all your mother's cooking pots from the kitchen. We had no idea what they were doing. We hoped they were going to be leaving in the morning, so we just ignored them. The next day after they had gone, we found all the pots up on the top floor placed around the walls. Each with a pile of human excrement inside."

I felt a wave of disgust sweep over me. We had often wondered why the smell coming from upstairs was like the soldiers had not cleaned themselves after using the toilet. Maybe the soldiers had been leaving their mess on the floor for years for nature to deal with it.

"The moralistic army used our cooking pots as lavatories," my father said, shaking his head. "They gathered everything, even empty bottles, sandbags and took them all with them, but, they used our cooking pots in this way and then left them behind deliberately as a souvenir."

He tried to make sense of what they had done, but could not. Did they somehow think that doing this would mock everything they knew he stood for?

"I am not angry," he said after a while. "But I am disappointed."

* * *

With the settlers gone, I finally had the chance to walk around K'far Darom, or at least what remained of it. It had been nearly a year since the settlers' evacuation. The walls of some of the houses still stood, and on a couple of them I saw drawings that had been made by the children. It is always the children who pay the heavy price for bad decisions made by their elders.

I had prayed every day for the settlers to go and leave us alone. Their presence had not been legal. And never could be. I was a Palestinian who had suffered at their hands. Yet, I was also human, and I ached for those children who were forced to leave what they had known as their home. It was not their fault that their parents had misled them and never told them the full story of whose land it is.

I smiled at the irony: they were forced to leave their homes, while I was desperate to leave mine. If I were to go to America, I needed my father's blessing. I now had a lot of explaining to do to him. I waited until he had finished his evening prayers and then went and sat in front of him. I told him the whole story about my application and my acceptance.

"Yaba, this school is one of the best prep schools in America, and they really want me to come," I said, and showed him the letter I had received. He was dubious.

"I am worried that this is a fake school, and you're just going to waste everyone's time, Yousef . . . So, you are telling me that you got yourself into a school through Google." Google was the new sensation of that time. He got up to walk away, but I did not stop. I followed him and tried saying all the wise things I had heard him say. I even tried some of the proverbs I had heard the elderly people repeat at their tea gatherings, like: "You should walk with the thief all the way to the end to find out if he is honest or not."

My father smiled, but kept moving and said nothing.

Sometimes when a guest came to the house, I would hear him say to my father, "He's too young. If he leaves now, he will become an American." I kept an ear out for that sort of thing. Whenever I heard words like that, I would pop out and tell my father he had a phone call. I had to ensure that no one had a chance to influence him and reinforce his fears. When my father went inside to take the "call," I would tell the guest my father would likely be on the phone for some time and, please, to come back later. I would grab the teacup from his hand and rush him to the gate before he even knew what was happening.

With the settlers and the soldiers finally gone, we once again had the freedom to welcome guests to our home without getting permission first. Every time a journalist came to talk with my father, he told them, "It has been agony not to have people here. It is our tradition."

Every day the soldiers had occupied our house, he had talked about having a grand party after they left to celebrate their departure. He wanted to invite all his Palestinian friends, of course, and also his international friends and journalists who had supported us while we were occupied.

"My hope for the next year," he kept saying, "is to be able to welcome the Israelis to my home, as friends, not as soldiers as before." He added, "When they were occupying my house, I am sure they were suffering exactly as I was suffering. All of us were suffering."

The seeming improvement of the situation between the Israelis and us only increased my faith that I was going to go to America. Believing that gave me an incredible boost. I had never felt so motivated about anything. I told myself that, not only was I going to go to America, I was going to achieve every dream I ever had.

The weeks of that summer weighed heavy while I waited for developments. With nothing else to do, I visited the site of the monastery that gave my city its name. It was the first time I had ever been there.

There was not much to see. When the monastery had been built, my city had already existed for centuries, long before war defined our history. With all that, you would think the monastery would look like a museum. But, no, it was just a few small abandoned rooms with vaulted ceilings lost in a jumble of other buildings, which boxed it in over the sixteen hundred years since its construction. Still, as I walked around them, I said to myself, "This is history. This is my history. I am from here."

I wondered if it would even exist in a few years. Everywhere I looked there was the gray wreckage of destruction. I feared that the monastery itself might vanish. I told myself that if I were ever going to have any chance at preserving the monastery and all the cities of Palestine I would need to go to America first. That was my rhetoric.

Finally, my father agreed.

When my mother phoned from Jordan, she whispered to me, "Don't you want to wait for me, Yousef? Can't you stay until I get back?"

It was the same tone she had used when she wanted me to stop watching *Tom and Jerry* and go to the shop for her groceries. However, this time I could not be persuaded. I explained to her with sadness, "I have to leave, Yama. I can't stay here. There is no future for me here. Please help my father understand. Please!"

If my mother had been at home, she would have been the one to help me prepare for my journey. Whenever someone in my family traveled, my mother would take them shopping for everything they needed. My father certainly did not have the patience that she had for that sort of thing. Since my mother was away, he was now the one who went shopping with me. He kept asking, "Are you sure you want to go away, Yousef? Are you sure this is what you want to do? It's so far away."

He gave me long lists of things to remember when we were apart. His favorites were, "Make it easy for people to love you," and "Try to smile at those who choose not to smile at you."

He showered me with wisdom, but all I did was nod my head

in agreement and look forward to being free from him and from everything around me.

* * *

It had become completely impossible for Palestinians to use Ben Gurion Airport in Tel Aviv, so it was decided that I should fly from Cairo. Even that would not be easy. We never knew when the border crossing to Egypt would be open, or when the Israelis would suddenly close it. All you could do was drive there and see. My father and I made seven trips to the crossing at Rafah without any luck.

To make things more complicated, the Israeli soldiers had set up checkpoints that split the Strip into sections, and they did so despite telling the whole world that they had given up Gaza. There was a checkpoint between our house and Rafah at Gush Katif, a former settlement. Sometimes the border at Rafah might be open, but the checkpoint at Gush Katif might be closed. Or Gush Katif could be open and the border closed. We never knew until we got there. It drove everybody living in the Strip crazy. Even if Gush Katif were open, it could sometimes take hours to get through.

We started hearing talk about a possible war with Lebanon, and I began to panic. If war broke out, the borders would be closed indefinitely. If I were to get to school in time for the autumn term, I had to be there by the middle of August. I kept praying that some angel would prevent a war, at least until I was free.

One depressing evening after we arrived home from the closed Rafah crossing, I heard my father shouting to come and listen to the television. The news anchor was reporting, "An Israeli soldier has been captured by Hamas."

I put my hands over my ears and started screaming, "Couldn't they just have waited a few more days?"

It was as if I were trying to build a castle in the middle of the sea. I went to my room to listen to my own little radio. It seemed clear that things were really bad, that Israel was very upset about what had happened, and that parts of Gaza were already on fire from rocket attacks. Whenever that kind of thing happened, it was the ordinary people who got punished.

The next afternoon, a taxi driver whose only route was to drive passengers to the borders blew his horn outside our house over and over until we came out to see what was happening.

"Abu Yazid, is your son still going to America?" he asked my father as he stepped out of the cab.

"Insh'Allah," my father replied. "If we can get him across the border. With this new fighting, though . . ." His voice trailed off.

"He may have a chance, if he leaves right now. I heard that a few people with medical problems will be allowed to cross into Egypt tomorrow. Maybe he can use his wound as a reason to get out with them."

After too long a silence, my father said, "A taxi is probably more likely to get past checkpoints than a private car."

"Yes," the driver agreed. "If we go right now, we can try to get beyond the checkpoint at Gush Katif this evening. It might get closed tomorrow. Your son can spend the night in Khan Younis." That was the town beyond the checkpoint where the driver would also stay. "Tomorrow morning, I can take him on to Rafah. Then he gets the bus to Cairo there."

From my house to Khan Younis is about seven miles. From Khan Younis to Rafah is another twelve miles. This is a two-day trip in Gaza.

I ran inside and grabbed my suitcase, which had been packed for days. It was heavy, because I kept thinking of more things to put in it. I knew that when I crossed the border this time, I would be unable to return home for several years. The thought was terrifying, but I kept it to myself.

"By the time I am back," I told myself, "we will have peace and things will be a lot better."

The driver had parked the taxi facing the newly built gate. I hugged my brothers and sisters, and then my father. For a moment,

I desperately wanted my mother to be there—but if she had been, it would have been very hard to say goodbye to her. I did not want anybody to see that. I was so ready to go, but it was very hard to leave knowing it might be a long time before I could see her again. The last time I had seen her was when I had left to go to the Friends School. Like so many other times before, my only hope of going forward was through the suppression of my feelings. It had to be done.

My father sounded worried as the boot of the taxi slammed, with my suitcase inside of it.

"Yousef, you'll be fine on your own tonight?" he asked.

"He can leave his luggage in the boot and either spend the night in the cab with me, or go to a hotel," the driver assured him. "During the night I'll drive around Khan Younis and pick up fares," he explained apologetically. "I have to make a living, you know."

As I jumped into the taxi, my father looked at me with those deep and serious eyes and said, "I will try to reach Rafah in the morning. I will see you there."

"Yaba, there is no need to come. I will be all right." As the taxi drove out the gate, I waved my arm out the window, but never looked back. I wanted to feel like I was moving under my own power after so many years of dreaming about it.

After all that, it took only an hour to get to Khan Younis. The taxi driver let me out in an area where there were a couple of small local hotels, and told me to meet him back there by 6:00 AM the next morning. There was a shop nearby, so I went there to phone my father and tell him I had already reached Khan Younis.

"Yes, Yaba, I will be fine for the night. No, I am not going to go anywhere. I will just stay in my room until it is time to go in the morning." My father sounded unconvinced.

A young man working in the shop overheard what I was saying and when I hung up he asked me, "Would you like to sleep at my house? It's not far from here and my mother will make us a delicious falafel and *ful* [fava beans] dinner." I was hesitant. I did not know him, and it seemed like an imposition.

"Come with me, don't worry about a hotel." He brought me some tea from a thermos he had behind the counter. "We will go in about an hour."

"Thanks," I said as I took the tea to a stool in the corner.

He lived in the Khan Younis refugee camp. I had no idea what to expect. He called his place a house, but it was no more than two tiny rooms. When we sat to eat, I hesitated to take any food as it was not up to my usual standards. Immediately I became disgusted with myself. They were sharing all they had with me. Anything they gave to me meant less for themselves. I thanked them and ate. It was simple, but actually it tasted very good.

He and I sat up talking most of the night about Palestine, about the world that so often seemed to forget us. He told me about how his grandparents were forced into exile before settling in the Strip.

"They thought the Arabs would come and take them back home," he said, not hiding his sarcasm. "They must have thought they would return home in five weeks, if it got really bad." Instead, it had been seven damn decades. He and his family had paid the real price by being labeled as "refugees" all their lives, suffocating their humanity.

The whole time he was speaking, I felt ashamed of myself. My family had suffered, but we still had our own land and had managed to hold on to it thanks to my father's courage. I told him I was on my way to America to go to school. He said I should be proud of all I had accomplished. Again, I felt uneasy. What had I accomplished? I was running away.

When it was time to go in the morning, he walked me to his shop and wished me luck.

The taxi picked me up a short while later. I left feeling as if he were a friend I had known all my life.

From Khan Younis, it took us three hours to get through all the checkpoints and go the dozen miles to reach Rafah. When we arrived, my father was already there. He had brought my two younger brothers with him. When I saw him I acted my usual independent self and said somewhat arrogantly, "What are you doing here? I told you I'd be fine." Secretly, though, I was very happy to see him.

He managed to get me into the line for the one bus that was heading to Cairo. As a sixteen-year-old, I never could have done that on my own. There were at least a hundred people with all their luggage pleading to get on. They yelled out their medical problems, hoping that would soften somebody's heart and get them a seat. My father kept phoning everybody he knew on the Palestinian side to get me onto that bus. He succeeded. Somebody grabbed my suitcase and hauled it into the bus. As I put my foot on the first step, I knew that I was very, very lucky. It was rumored that after this bus went through, the border would be closed again for several months.

As soon as the bus was filled, I thought we would leave. Instead, we sat there for two full days waiting to be allowed through the checkpoint. The whole time, my father remained beside the bus, sometimes leaning against his car, sometimes sitting in it, always talking to me through our open windows. I had to hang half of my body out the window so I could hear him.

Occasionally, he napped briefly in the car while I talked with my brothers. Being at the border was exciting for them. They enjoyed buying food from vendors, and seeing other kids their age selling ice cream, cold drinks, and tea from trays, suspended from their necks by straps. My brothers laughed at how the tea sellers would stand and wait for their customers to finish their tea and then take back the cup and refill it for the next customer without washing it. They cracked up about things they had heard people say as they were milling around and waiting to cross the border. There were hundreds of them.

At night, while I slept, my father stayed awake reading, or praying on the road next to the car. My brothers slept in the back seat.

The bus had no toilet. So, I did not dare eat anything. I could not risk leaving the bus to find a toilet. Either somebody might grab my seat, or the bus might leave while I was away and I would be stuck in Gaza. It was summer and very hot, so I had to drink. My father passed cold drinks to me through the window. After dark and when all the other passengers on board were asleep, I peed into an empty Coke bottle, and passed it through the window to my father outside. I was not the only one doing this. I was in the last seat at the back of the bus, which was good, since that gave me a little privacy.

Then suddenly, around noon on the third day, without warning, the gate opened and the bus began driving slowly forward. My father had been sitting in his car, talking with me when he heard the bus engine start to rumble. He jumped out of the car and started jogging along beside the bus.

"Yousef, be nice over there," he shouted at me through my window. "Make sure to study hard. Good marks will open any door you want to open."

As the bus picked up speed, he started running, with my little brothers racing along behind him waving and smiling. The bus began moving faster and faster. He kept pace with it, jumping over things to maintain his view of me. I kneeled on my seat and slammed my palms against the back window, staring at my father as the bus pulled away from him.

He kept running, waving his hands and calling to me, "*Ella al-amam, Yousef* [Forward, Yousef]!" It was the last thing I heard him say. I was leaving and I had not even managed to properly say goodbye. I continued to stare at him until the gate closed and he was out of sight.

"I'll see you again soon, Yaba," I said. "I'll make you proud." When my mobile phone rang, though, and it was him, all I could say was, "Yaba, I told you I'd call you when I got to Cairo. Let me call you when I get to Cairo."

Part Five

LEGACY

"Forgive and forget. God loves those who
honor him by forgiving."
—The Holy Quran, Surat Al-Ma'ida (5:13)

19

America

My father had asked my brother in Germany to buy my plane ticket and he had chosen Air France. I had to change planes in Paris, and it proved to be a bit overwhelming. I did not speak a word of French and had no idea where to go or what I was supposed to do. Even worse, when I got to the immigration area, I was told that there was a problem with my passport and I would have to wait. I had no idea why.

All I can remember is standing in front of the desk and pleading with the lady, "I'm not going to be a problem. I love peace. I want to make peace." I was also worried about my luggage and the one thousand dollars my father had given me to give to the school. I kept counting and recounting that money, afraid I might have lost it or someone had stolen it. Finally, just as the plane was boarding, they gave me back my passport and directed me to my gate.

The plane took forever to cross the Atlantic. I kept watching its image on the screen in front of my seat, but it never seemed to move. Once I even called the stewardess and told her in my most adult voice, "There is something wrong with this screen. The plane isn't moving." She only laughed and said, "You're funny."

When we landed in Atlanta, the immigration officers already knew everything there was to know about me and where I was going and why. I could not imagine how they knew all that, and, honestly, I was impressed. I changed planes for Utah, where I arrived late at night. I was glad to see a guy holding a sign with my name on it. The first thing I noticed was that he had an earring in one ear and wore his cap backward.

197

When I had dreamed of coming to America, I had presumed I could leave all the nagging memories behind me. I thought I could forget the pain and all the war I had witnessed. I told myself that once I got to America my life would be normal again. Yet there I was, in a safe and wonderful place, and at night I could not sleep. Every night I would have the same terrible dreams—dreams of being in my house, dreams of being awakened by soldiers, dreams of being shot.

In my most frequent dream, I was giving a speech in the chapel at the Friends School in Ramallah and could see a man standing off to one side silently staring at me. I would try to continue speaking, but I somehow knew that if he got close enough he was going to shoot me in the chest. I had to ensure I got my message across before that happened. I would jump out of bed and try to wake whoever was my roommate at the time.

I worried that my father would make fun of my dreams if I told him about them. It was his way of toughening me up. Sometimes when he phoned he would ask, with a little laugh, "Do you miss me, Yousef?" I would yearn to say, "Of course I miss you, Yaba. I miss you so much!" But I pretended I had not heard him.

My father had often made fun of me, especially when he thought I had done something stupid, but usually he got angry first. Like the time I tried to bribe my religion teacher with vegetables from the farm, hoping he would pass me even though I had not memorized the *Surah* he had assigned. My father never let me forget that.

As my time in America passed, things back in Gaza were worsening. On top of the huge problems that Palestinians had to endure daily, they were now fighting with each other about how best to deal with Israel. It broke my heart into pieces and began to eat away at my idealism and my hope for the future.

When I had first come to the US, I called home a lot. As things became increasingly difficult there, I called only about once a week, just to make sure everyone was still alive. It was too hard to hear about all the problems at home and to not be able to do anything about them. My brothers would also call me from Germany and my sweet older sister even sent me a Kodak camera for my birthday.

"Take pictures, Yusfi, I want to see you," she wrote. It seemed so pointless though. Our lives were becoming so different, how could a picture make up for the distance?

I told myself, "My father did not want to leave his home, but I had to. I have to find a place where I can build a future in peace."

To me, living in peace meant living as I had lived as a child, with no fear. It meant a life filled with love, family, the farm, the school, the olive trees, the orange trees, the palm trees, and the goats and chickens.

Now, when I was feeling bad about myself, those old memories helped me feel better. I knew from experience that when I had a plan, I could make it happen. I had learned to walk again, I had made it to America, and I would continue to find a way to achieve my goals.

In late 2008, three years after I arrived in America, Israel invaded the Strip. I kept calling and calling, but Gaza had been shut off from all contact with the outside world. I was choked with worries about my family. I was unable to talk with them for a whole week. It was strange to be on the other side of the ocean and watch my homeland being engulfed in hatred and fire. I had thought that by being in America I would be free from worry about war, but knowing my family was again living under threat of war made me feel like I had never left. And worse was understanding that the Israeli occupation of Gaza never truly ended.

I knew that I did not even dare go for a visit. It was too big a risk to take. Of all the terrible things that could happen to me, the worst would be to go home and not be allowed to leave again. I had been lucky to leave the first time and I could not guarantee I would be that lucky again.

Even though I was in America, my head was often in a bad place. It was hard having the bullet always there, pressing on the nerves beside my spine. My brain was constantly telling me that I was in pain, especially when I tried to sleep. I kept switching positions, trying to find some way to lie that did not hurt. The frustration was as intense as the pain.

If I slept too long, it bothered me. If the weather got too cold, it bothered me. It affected my legs as well as my back, and it was even starting to affect my sanity. There were times I wanted to scream out loud, pound on the wall, break the windows, anything to release the tension. There would be times when I could barely stop myself from ripping off my clothes. I somehow felt that if I could just get my clothes off, maybe I could reach inside my body and pull the damn pain out.

It was hard to control my emotions, and tears would start to come to my eyes. There were days when I just wanted to break down and sob, but I would fight that feeling.

I began to obsess about the soldier who had shot me, wondering why he had done so and what lesson we were both supposed to learn from it. That single shot had changed my whole life and I wondered if it had changed his. I wondered about his name, his age, and what he was doing with his life. I fantasized that he was probably living some great life on a beach in Thailand. I had heard that Israeli soldiers often went on holiday there after completing military duty.

Sometimes I did allow myself to have thoughts of revenge, but I knew I had to deal better with my pain. I had to remember what my father had tried so hard to teach me. I had to remember that the moment I let hatred flow into my heart, my life would become meaningless. I could not let that happen.

As time passed, my conversations with my father were getting better. When we spoke, I forgot about my troubles and my fears and, instead, focused on all the interesting meetings I had attended and all the fascinating people I had met. I loved telling him about my invitations to speak, and he loved hearing about them.

I told him how Adam, whom I had met at Seeds of Peace camp and who was a member of Group F, called me one day when I was at Wasatch to invite me to Boston to speak to his synagogue.

"My parents will fly you both ways and it would be great to have you here. Everyone will want to hear your story. I know they will be touched by what you have to say." I had never spoken to a Jewish audience and the thought was intimidating, but it sounded like an amazing opportunity. Besides, I would finally get to experience Boston, my dream city, and not just from the Seeds of Peace bus as it drove through.

It was autumn when I arrived, and all the trees were a brilliant red. It was so beautiful, and I was so happy to be there. Adam's parents hosted a dinner in my honor to raise some money for me. When I arrived at the synagogue where I was to speak, there were a thousand people there. I was stunned. Outside of school, I had only spoken in public once before and that had been at camp. How could I speak to a thousand people?

The truth is, though, I loved telling my story. I can still see the way everyone listened to my words. I can still feel how everyone's eyes were focused on me and on what I was saying. For the first time in my life, I felt powerful. With every word, it became easier for me to speak. It was easy because I was speaking words I had learned from my father. It was easy because I was in love with the message I was sharing. It was a message I hoped to always preach, no matter how many bad things might happen to me because of it. I hoped I would never forget that my enemies were also human. I hoped I would never surrender my heart to fear and anger. Suddenly, I realized that the bullet in my back could serve a purpose and need not hold me back.

I loved it when I put down the microphone and everyone stood up and clapped for me. I loved standing in line and greeting people and shaking their hands and answering their questions. I loved the fact that they were so interested in what I had to say and wanted to know more. And I loved it when I later learned they had sent three thousand dollars to my school to help pay my tuition fees. That was not nearly enough to pay off my debt, but it certainly helped and gave me a whole new idea: perhaps, if I spoke at more places, I could come up with the rest of the money I owed. I had seen famous people on TV raise money for their causes, so maybe someone who was not a celebrity could do that, too. A little voice inside of me began to chant, "Don't give up, Yousef. Don't give up. Be calm and things will work out for you."

My father was very excited when I told him about the Boston speech, but he soon returned to his old worries about me.

"But Yaba, don't you believe in me?"

"Yes, I believe in you, Yousef, but sometimes you make it hard for me to trust you."

My father's lack of trust in me had a long history. Whenever I had been punished as a child, it had usually been because I had been too scared to tell the truth. I was often scared to tell the truth before giving my own plan a chance to come to life and prove that I knew what I was doing. I wanted my father to trust my judgment, and to be proud of me.

Perhaps it was time to break down and tell my father about the money problems I was having paying for school and housing and all the other things that do not come free. He had warned me over and over that if things did not go as I had hoped, he would not be able to help me. On those increasingly rare occasions when I now called home, he asked, "Yousef, are you hungry? Do you have any money?" The more he asked, the more I hesitated to answer.

I could not bear to tell him that I had never managed to fully pay my tuition at Wasatch Academy, but that the kind headmaster there had waived everything I owed as I was graduating. I could not tell him that though I was given a scholarship to a very fine small university in Berea, Kentucky, I transferred out after one term to go to Boston, the city of my dreams. I could not tell him how I was about to be forced to leave Suffolk University in Boston when the scholarship I had been given was being withdrawn because I had not kept up my marks.

Or how I walked into the historic Old Granary Burying Ground across the street from Suffolk University and talked to the graves. "Hi, I am Yousef from Palestine. Please tell me what it's like on the other side." I threw myself on the grass in sorrow and started weeping. Some guy saw me and yelled, "Get the hell out of there."

Even when I could convince myself that my problems would work themselves out, I could sense that my father was really worrying about me. He said things like, "Yousef, something tells me you're not doing well. I'm really worried about you, Mr. Yousef." Sometimes he added, "You're not telling me the truth. I know you, Yousef!" I told him he was crazy, and changed the subject, but he knew he was right.

A Syrian professor at Suffolk University took me in for a while and helped me get on my feet. Some Palestinian Americans living in Boston heard about me and collected some money for me. They helped me find a space to live and got me a job at a local restaurant that was owned by a Palestinian man. He said he had to pay me under the table and could only afford seven dollars and fifty cents an hour. I could tell there was no point in trying to bargain for more. I needed the money and would settle for whatever he offered.

I discovered that I loved working as a chef, especially cutting and grilling the lamb and chicken. I even learned how to make rice. After only a couple of weeks, though, the owner and I started fighting over portion control. He criticized me for putting too much food on the plates and I criticized him back for not caring about his customers. He kept saying he was teaching me about restaurant management. I wanted to tell him, "I'm not here to learn how to run a restaurant. I'm here because I need cash in order to live." At least as long as I worked there, I could truthfully say to my father that I had enough to eat.

I knew if my father found out how bad things really were, he would sell his land to support me, and I would have to live with that shame forever. He had told me so many times, "The land shall always stay as it is." He had managed to keep his land safe from an entire army. How could I live with myself knowing that he had been forced to sell it just to send me to school in America?

I had put myself in a world that was different from anything I had ever known. This was a world that did not care if I had nightmares or endless pain in my back and legs. All I had to offer this world were my little words about the need for peace. All I had to give was my conviction that there would be no peace in the Middle East until the people of Palestine and Israel learned to live comfortably with one another. I had no assurance that any of that would matter to anyone else.

20

My Father

One night as I was visiting Hussain—a Saudi friend I had met at Suffolk University and who lived across the river from Boston in Cambridge, Massachusetts—my brother, Yazan, phoned from Germany. I was not expecting a call from him and was surprised to hear his voice as I sat on Hussain's white couch.

"Yousef, can you hear me?" he said, without either a "hi" or "how are you?"

"Listen, Yousef," he stated, with an unusual urgency to his voice. "I need to tell you something."

"Well, OK," I responded impatiently, "Go ahead, tell me."

He continued, "You need to be very strong, Yousef. You are a man now."

Still puzzled, I urged him again, "OK, just tell me already."

Finally, he managed to say it. "Yousef, our father just passed and I want you to know that I'm thinking about you. I'm sure you will stay strong, OK?" With those words, his voice choked and he hung up.

I could neither breathe nor speak for a moment. I felt as I had when I was shot—paralyzed. I covered my mouth with my hands and ran outside. I ran as though I were escaping from time, from reality. It did not matter that I had no more breath; my legs just kept running. Hussain raced to his car and was now driving behind me, shouting, "Stop, Yousef, stop. Where are you going, brother?" I just kept running and running, until I finally collapsed on the Longfellow Bridge over the Charles River.

Hussain jumped out of his car and tried to get me to stand, but I could not bear my weight. My body and my mind were numb and

I was screaming as I had never screamed before. I could not believe what I had just heard. I had never really allowed myself to contemplate life without my father. As much as I had challenged him and rebelled against his demands and his way of life, he had always been my anchor. The last time I had seen him had been at the border on my way to Cairo and the US. That would be my last time ever to be with him on this Earth.

"*Ella al-amam, Yousef!*" It kept ringing in my ears. "Forward, Yousef!"

Hussain took me back to his flat. When I had calmed down, I called my mother. I could hear her speaking through her tears, trying desperately to reassure me, "Yousef, your father left in peace. He died lying where he wanted to be, Yousef. That is a good thing, Yousef." She told me he had been found lying on his land, the land he had spent his whole life defending, defending with words of peace.

She said it had been sudden, probably a heart attack, but that did not soothe me. She kept repeating, "He died in peace, Yousef. Let that comfort you." As she spoke, I could hear her strength and knew I had to find my own. I felt so sorry for my mother, who had borne so much and never complained. I wished I could take away her sadness, but I was too helpless and defeated myself.

"I will come for the funeral," I told her, as if that was going to make everything better.

"No, Yousef," she insisted. "Your father wouldn't want you to leave school behind. Don't come, the borders aren't open. Don't risk it, my dear son."

"Yama, I have to come." But I did not go.

I spoke to my family every day, desperate to know every small thing that was happening there. My younger brother Mohammed Salah gave me the details that I clung to, wanting so much to be a part of them myself. He told me how, the night before, my father's sisters had come to the house. Ramadan was nearly over. The approaching special occasion, Eid, was filling everyone with joy.

"He spent the night laughing and giggling with his sisters until they could not laugh anymore," Mohammed Salah said. "Every time his sisters got up to head home, he persuaded them to stay a little longer and they easily gave in." He had spent the day planting cauliflower and cabbage seedlings, and was enjoying relaxing with the family. He intended to finish the planting the next day.

In the morning, my father woke first and tried to get everyone else up to eat before *Fajir*, morning prayer time, after which a new day of fasting would begin. My mother was still tired after serving a large *Iftar* supper to all the guests the night before. She was moving slowly and let my father take charge. My father went to Mohammed Salah's room and turned on the light. In a sleepy voice, my brother complained that the light was too bright.

"But it is almost time, Yaba," my father said to Mohammed Salah as he turned the light back off. Those were the last words he said on this Earth.

After he had finished eating by himself, my father went outside around 6:00 AM to pray and get the irrigation system ready before the workers arrived. He always disliked being late or keeping anyone waiting.

Inside his room, Mohammed Salah was attempting to get just a little more sleep before he joined the planting party, but was awakened again by loud knocks that he realized were the workers at the gate. He could not understand why they kept on banging and why my father had not let them in.

As he was running out of the house toward the gate, he stopped short as he reached the guava tree. My father was lying on the ground beneath it, looking up at him. Mohammed Salah went over toward him asking, "Yaba? Yaba?" My father said nothing. Mohammed knelt beside him. "Yaba? Can you hear me?" No answer.

Mohammed Salah instinctively understood what had happened.

"I got up and made a few steps backward to go tell my mother, but then I stopped and walked back to my father. I did not want to leave him alone. So I knelt down next to him, but I could tell that his soul had already moved beyond this earth."

After a few moments he accepted that he had to go tell my mother. He broke the news to her as she was getting out of bed. She hugged him as if he were her father and sobbed in pain and outrage, then went with him out into the garden. When she saw my father she began talking to him and begging him to respond to her, "Ya Khalil? Ya Khalil?"

They carried him inside the house to the "prison room" where we had been imprisoned by the soldiers. My younger sisters prepared a comfortable setting for him with *freash* and pillows and covered him with a blanket.

Mohammed Salah ran outside to find someone with a car. When he opened the gate, the workers who had been banging on it started complaining that he was slow to let them in. He ran right past them and stopped a man driving a Volkswagen van. Without a second thought, the driver turned around and drove to the house when he heard that something had happened to my father. He and Mohammed Salah carried my father to the car, then drove him to the hospital in Deir, but there was no hope.

As word spread through Deir el-Balah, people rushed to the hospital demanding to see my father. Mohammed Salah tried to keep them away but people refused to go. Even the staff at the hospital crowded into the room where he had been brought, while others filled the hallway, all in shock that Khalil Bashir had passed away.

They drove his body back to the house where a sheikh had arrived and was waiting to pray for him and prepare his body for burial. People were streaming into the house. My history teacher, a religious man who dearly loved my father, insisted on coming to see him and say his goodbyes. He and another teacher were the first to arrive. They each held his head in turn and kissed him gently, their voices filled with mourning. His sisters started weeping and crying over his head when they saw him lying silently.

My mother at first thought it would be wiser not to inform Yazid, who on that day was scheduled to have his final exam—the one that would decide if he was to become a doctor or not. My mother could not hold herself back, however, and phoned him to tell him what had happened, as she had to.

The mosque that had been chosen for the funeral prayers had to be changed because it was too small. The prayers were moved to another mosque that had been built by my mother's father, because it was larger. Even then, it still was not large enough. Crowds stretched out to the street when they found no place to pray inside. Everyone was there and everyone was pushing hard to get to carry his coffin on the march to the cemetery.

Usually the police do not allow funerals to shut down the streets, but there were so many people that they had to close several roads to make it safe for people to walk. Midway to the cemetery some men attempted to lower his coffin from the shoulders of those carrying it so that they could say their prayers and goodbyes. My brother had to push them away over and over.

Mohammed Salah told me about a man who was at the market shopping for vegetables. "When he heard the news that the coffin that was passing held Khalil Bashir, he threw his bag to the ground and stepped over his vegetables to run and join the crowd."

After my father was laid to rest, the hundreds who had accompanied him to the cemetery joined an even larger crowd at our house. Mohammed Salah said, "It was like I was seeing a long, long snake made of people and cars, Yousef." The Minister of Education gave a eulogy in my father's honor at the house, as more people continued to arrive. No less than two thousand people were there.

Since it was about to be Eid, my mother wanted to follow the tradition of not holding a funeral on a day of celebration in Islam. She felt she should send everyone away, but the people would not leave.

Between my endless phone calls home, I thought of the last time I saw my father on the border. I decided that my life must be dedicated to him. My father will live on in me and that is how it must be. I will do everything in my power to embrace him and open the doors of my rebellious soul and become one with his.

There then followed an impossible time for me, as I struggled with the unresolved issues of many years. I had to find a way to solve my problems. One day while shaving I looked at myself in the mirror with anger and hatred. "Wake up, Yousef," I shouted. "Wake up, you stupid guy." I punched the mirror, then went and sat on the edge of the bathtub observing my bleeding hand.

Slowly, some relief began to seep into my consciousness. Just when I thought my father was gone from me and just when I thought I could no longer talk to him, I began to sense that he was standing right in front of me. I began talking to him through my inner voice. I found we could have a conversation, his voice in my heart and my voice in my mind. I felt embarrassed at first, but with time I found the courage to tell him my story and all that had happened to me. I wanted to tell him the truth. I needed to tell him the truth, the whole truth.

When people in Boston learned of my loss, they were amazingly kind. Even strangers reacted as though they were close family. Everyone was calling me. Everyone seemed to want to be there for me, and I doubt I could have maintained my sanity without them. I hated having people feel sorry for me, but I needed their kindness.

In those dreams that happen when you are not quite asleep and not quite awake, my father came to visit me, and even brought my mother with him. Her voice swept through my furniture-less flat, wondering what she should cook for me and if I had detergent for my laundry. I did not believe the illusion and kept on struggling to fall asleep, but my beautiful smiling father, dressed completely in white, sat on the right edge of my mattress.

"Would you like to have tea with me?" he offered with his usual enthusiasm. "Do not be afraid, everything is going to be fine, you are going to be fine," he added, before standing up to go argue with my mother about something she had or had not done, while I made the sad mistake of waking up.

My time with him would now have to come from my memories. I now had to absorb into my own being the fullness of his words: "What happened to me makes me believe even more in peace." I had to trust that, from now on, he would be with me on my journey and I had to ensure that my journey would be one that pleases him.

Though I had protected my father from my truth while he was alive, I was sure that now, in the spirit world, he knew everything. He could see everything and I could no longer hide anything. I had to make my life work, because he would know if I did not. I could never allow myself to give up. I had inherited my father's mission and it was now my job to help the world learn what it meant to resist in the name of peace. His soul mingled with mine as though, by his death, he and I had become one.

From now on, the words of my father shall become my own. I shall spread them the same way he spread seeds on his beloved land and brought them to life. I shall speak for the two of us.

To be continued . . .

A Letter to the Soldier Who Shot Me

Dear Cousin,

Why?

I have many questions to ask you, but the most insistent one is why. Why did you shoot me? What made you fire at a skinny fifteen-year-old kid, who was standing next to his father, waving goodbye to some visitors?

What did you feel as you squeezed your trigger? What did you think as you saw me crumple to the ground? Did you laugh? Did you curse?

Those visitors had been there with your commander's permission. They left as soon as one of the watchtower soldiers (maybe you?) told them through a loudspeaker that they had to leave.

They were from the United Nations and were traveling in a vehicle with "UN" painted on the side. For the fifteen minutes that they had been at our house, they had been speaking with my father in front of our house where you could watch them. What was it about them—or us—that filled you with so much anger?

Do you ever wonder what happened to me?

Your superiors came to me in the hospital. They would not tell me your name. They apologized for what you did. You never came.

Did they ever tell you that I had a miraculous recovery, thanks to Jewish doctors, Jewish nurses, and Jewish therapists in Tel Aviv?

Many times I have thought about you. Were you born in Israel, or were you someone who came from Brooklyn looking to fight? What did you tell the other soldiers that night, after you shot me? Did you boast? Did you cry? Were you frightened?

Are you married now? Do you have children? What do you tell

them about me? About why you shot me? About what you have felt since that day? Or have you completely forgotten about me?

It is not possible for me to forget you. The three pieces of your bullet embedded in my back, next to my spine, make me think of you every day. I wanted to hate you, but a miracle happened. No, not the miracle that I can walk again. Another miracle. One that was shown to me through my father's commitment to peace, my mother's unfathomable love, and the doctors and nurses who tended to me with the deepest compassion. It is the miracle of forgiveness.

Without your bullet, I might never have understood forgiveness. You were created by the same God who created me. You have the same humanity as I have. You are part of the same family as I am.

I forgive you, my cousin.

In peace,

—Yousef K. B.

The Words of My Father

My father expressed his commitment to peace in numerous interviews with the media from all parts of the world. Here, in his own words, is a small sampling of what he said.

"I witnessed three wars and two Intifadas and now I'm thinking of my children's future. I don't want them to see war, and the only way to prevent that is to overcome the mountain of suspicion. We are destined to live together with the Israelis. We have to change our mentality. If we let our wounded memory guide our future steps we will have only pain."

—*The Guardian*, 3 October 2005

"We are destined to live together in this land. We have to share it. Let us share it."

—*The Philadelphia Inquirer*, 15 August 2005

"Because I do not feel any sense of hatred toward the other side, and have already forgiven everything, and because of my deep belief in peace, I have managed to stay in my house and remain sane. . . . It is the politicians that block the mutual understanding between our two people. I hope for the day when this mountain of ice that is our disagreement will melt.

"[My house] is like a metaphor for our people. The Palestinians gave up more than two thirds of Palestine and now, like my house, it is the playground of our cousins, the Israelis. The Israelis have to go back to where they were before 1967, let the Palestinians establish our own independent state with East Jerusalem as the capital. Understand that if you crush your enemy, this is not peace, as everyone must be happy.

"Please, please, convey the following to my neighbors the Israelis. Please, enough hatred. Enough war. Let us look optimistically to the future and give our children the chance to live in peace."
—*Sabotage Times*, 16 October 2010

"Suppose I think of taking revenge. What is the result? What is the outcome? I do believe that to forgive gives room for changing the mentality."
—*CNN*, 13 August 2005

"There is a chance to coexist with the Israelis. I sometimes read this in the eyes of the soldiers who come and imprison my family and me in only one room. I read in their eyes that they behave professionally, but they are not willing, they are not satisfied with what they are doing against me."
—*BBC*, 13 May 2005

"I am still asking the Israelis to shake hands with me. I challenge them to invite me to visit their houses, and I will invite them to visit mine. But up to this moment the only answer they gave me was to shoot Yousef."
—*The Guardian*, 2 March 2004

"They will never succeed in making me hate them. We are better than that. To hate is inhuman. In spite of everything, I love them and reach out the hand of peace. We are destined to live together. Nobody will expel anybody else. We are all children of Abraham.

"I will never lose my optimism. There are people on both sides who want peace. The real battle is not between Israel and Palestine, but between those who want to coexist and those who dream of expelling or killing the other side.

"God's protection was more powerful than the hatred of the bullets, for it prevented death from taking Yousef. We are all human, and I pray I am the last father on either side who suffers, and Yousef will be the last son who gets hurt."

—*The Independent*, 5 March 2004

Glossary

Aboi	How one refers to one's father when he is not present
Al-Aqsa	The third holiest site in Islam, located in Jerusalem
Allahu Akbar	God is great
Al Rasheed Road	A road that runs the length of the Gaza Strip along the coast
Amr Diab	Popular Egyptian singer
Amreeka	"America" in Arabic
Azan	Islamic call to prayer
Dayenu	Traditional Jewish song sung at Passover
Eid	Muslim Holidays. There are two: Eid Al-Fitr which comes at the end of the holy month of Ramadan and lasts for three days, and Eid Al-Adha which comes in the twelfth month of the Islamic calendar and lasts for four days
El Yahood	The Jews
Erez Crossing	Border crossing located at the north of the Gaza strip and controlled by Israel
Fajir	The first of the five daily prayers offered by Muslims; during Ramadan it marks the time to start fasting
Falafel	A deep-fried ball made from ground chickpeas, fava beans, or both
Fanous	Lamp or light used in Ramadan

Fattoush	Bread salad made from toasted or fried pieces of pita bread combined with mixed greens and other vegetables
Feseekh	Traditional Egyptian fish dish made of fermented salted and dried fish
Freash	A slang term for the flat cushions used for seating on the floor
Gatayef	An Arab dessert commonly served during the month of Ramadan
Goa'uld	A symbiotic race of ancient astronauts from the television franchise *Stargate*
Gush Katif	Former Israeli settlement in the Gaza Strip
Halal	Meat from animals slaughtered according to Islamic dietary laws
Hassidim	Ultra-orthodox Jews
Hoy hoy	Slang term used to downgrade someone in a teasing way
Hummus	A dip or spread made from cooked and mashed chickpeas blended with tahini, olive oil, lemon juice, salt, and garlic
Iftar	The evening meal when Muslims end their daily Ramadan fast at sunset
Insh'Allah	"God willing" in Arabic
Jalabiyyah	Traditional Arab garment
Kaboria	A male haircut style that circles around the head leaving the top part uncut while the lower part is completely shaved
Kadima	"Let's go" in Hebrew
Kalandia	Checkpoint near Ramallah in the West Bank, controlled by Israel
Kanoon	Brass brazier
Kazem	A well-known ice cream maker in Gaza
Keffiyeh	Traditional Middle Eastern headdress
Khan Younis	A city in the Gaza Strip

Kishk	A delicacy made from drained yogurt which is then dried
Kosher	Food prepared according to Jewish dietary laws
Laban	Yogurt
Mamoul	Small shortbread pastries filled with dates, pistachios, figs, walnuts, or almonds
Maqlouba	Traditional dish that includes meat, rice, and fried vegetables placed in a pot which is then flipped upside down
Mashe	Arabic street word for "OK?" or "sounds good?"
Mazboot	"Right?" in Arabic slang
Merkava	Israeli-made tank
Nuseirat	A city in the Gaza Strip
Oskoot	"Shut up" in Hebrew slang
Rafah	Southern city of the Gaza Strip
Rafah Crossing	Southern exit from the Gaza Strip to Egypt and the world
Salah al-Din Highway	The main highway that runs the length of the Gaza Strip and crosses into Israel in the north at Erez and into Egypt in the south at Rafah
Shahada	The Islamic Creed, one of the Five Pillars of Islam. "There is no god but Allah; Muhammad is the Messenger of Allah."
Sheikh	A title that commonly designates the chief or head of a tribe, family, or village, who inherited the title from his father; this also often serves as a title for Islamic clerics
Sitie	Term used to refer to a grandmother
Sufra	Traditionally low table, tray, or cloth used for dining

GLOSSARY

Suhoor	The meal consumed before sunrise during Ramadan
Surah	A verse from the Holy Quran
Tabeekh	A stew
Tabbouleh	A salad made of tomatoes, finely chopped parsley, mint, bulgur, and onion, seasoned with olive oil, lemon juice, and salt
TOEFL	Exam testing English language ability of non-native speakers
Yaba	Term used to address one's father, which can also be used by a father to address his son
Yama	Term used to address one's mother

Acknowledgments

I believe that God is the All Forgiving Most Merciful and it is man who causes violence to erupt. I am thankful to God to have the opportunity to share *The Words of My Father*. I have written this book because I believe there are ways to fight for what is right other than with violence.

I write because it allows me to fight for my freedom and for a life of justice and dignity. I write for peace and forgiveness, although my home is under siege. I write because I think I have finally understood Rumi when he wrote that there is no love greater than a love without a lover, and my commitment to peace has been such a love affair without a lover.

I have been blessed by God to live in Boston, which to me is the place of love, though I have lived a life of occupation, too, for I am from Palestine and it is my motherland.

I write for my Father and Mother
I write for my Sisters and Brothers
I write for my Enemies and Friends
I write for you and all the others

I have watched the tales of war grow not only within my homeland but far, far beyond. More than ever, I firmly believe that there is no way forward but to make peace in the Holy Land.

For those who helped turn my story into a book, I am deeply grateful.

My deepest appreciation goes to Professor Eleanor Macklin, who worked tirelessly with me as a mentor and a friend to help shape what I had written into a story worth reading.

Profound thanks to Stephen Landrigan for his skills and sensitivity as an editor and for his unfailing belief that the words of my father need to be heard everywhere.

Also, I thank:

John Schappi, John H., Medea B., Mr. Loftin, Michael R., Henry O., Leonard R., Peter E., Aaron S., Rudolf W., Bill B., Wendy T., Deeann G., Barbara V., Smitty S., Munir A., Michael B., Dr. Brezener, Caroline and Pat H., Diane J. and David S., Erv and Kate L., Janice K. and Michael A., Bobbie and Tom G., Katherine B. and Gene W., Tim Lisa L., Orphee and Glaydess D., Mardge and Gordon C., Bob and Karen S., Doug and Caroline H., Bob M. and Steve B., Judy and Jeff W., Colin and Virginia S., Jennifer and Eric M., David W., Tom R., Abraham G., Paige G., Leslie A.L., Eric K., Eva G., Daniel D., Dalie J., Matthew D., Tim R., Daniel T., Joseph B., Sally S., Jerry M., Mr. Khaldi, Mr. Abul-Roos, Mrs. Johnson, Mrs. Turner, Mrs. Austin. Mr. Austin, Elaine B., Marybeth K., Prof. Rufin, Prof. King, Prof. Sullivan, Prof. Jones, Prof. Eijmberts, Prof. Turam, Prof. Uzdella, Prof. Kohanteb, Prof. Mehozay, Prof. Lemon, Prof. Douglas, Prof. Norris, Lana C, Michele C.D., Mr. Bey. Fessial M., Andrey K., Daniel A., Tariq S., Kaamil R., Hussain and Saif A.L., Sam R., Adam S., Ben L., Jonathan K., Mohammad Salah., Andy M., Lily A., Nikos P., Antoine L., Mohammad Amine H., James W., Simon S., Chris H., William E., Tom G., Mira R., Daniel S., Elena B., Courtney P., Aaron D.M., Robin W., Ian D., Colin D., Amy W., Laurie D. and Joe S., Lynne T., Laurie and Robert W., Brenda G., Leila S., and Isabella J.

The Friends Boys School, Ramallah, and The Quaker Friends House, Boston. Wasatch Academy, Berea College, Suffolk University, Bunker Hill Community College, Northeastern University, Brandeis University, the Herbert Scoville Jr. Peace Fellowship, the Offic of Congressman Gerry Connolly, the Offic of Senator Bernie Sanders, Seeds of Peace, Ropes & Gray LLP, and the Pair Project.

You know who you are. You are all part of my story.

About the Author

YOUSEF BASHIR is a Palestinian American from the Gaza Strip and the son of Khalil Bashir, a highly respected educator. Still suffering the effects of a near-catastrophic injury at the hands of an anonymous Israel Defense Forces soldier, Yousef made his way to the United States, where he earned a BA in International Affairs from Northeastern University and an MA in Coexistence and Conflict from Brandeis University. Now living in Washington, DC, he has worked on Capitol Hill and served as a member of the Palestinian Diplomatic Delegation to the United States. Yousef is an accomplished author, a vigorous advocate of Israeli-Palestinian peace, and a much sought-after public speaker.